Balcombe and Beyond
The UK's Frack Free Movement

Martin D. Dale
Foreword by James Bolam and Sue Jameson

Also by the Author:
Pulborough's Fallen: The Casualties of Two World Wars
The First World War Roll of Honour Vols. 1 – 17

Author Contributions:
Bringing it Home National Poetry Anthology
Great War Britain: West Sussex Remembered 1914-18

LEGAL NOTICE

Due to the nature of this book, care has been taken to ensure that photos containing people have been avoided where possible. Where a person is visible, the photos were selected so as to ensure the person is not identifiable in order to protect their privacy and identity. Photos containing children or Police Officers have been avoided.
Photos of protest activity have been restricted to those not depicting groups of people or identifiable individuals.

To the Author's best knowledge, all photos have been taken from public land and/or public rights of way.

First Edition

Copyright © 2015 by Martin D. Dale unless otherwise stated.

The moral right of the author has been asserted.

All rights reserved.
No part of this publication may be reproduced, stored in a retrieval system, or transmitted in any form or by any means, without the prior written permission of the author, nor circulated in any form other than that in which it is published without the express written permission of the author.

ISBN: 1505265738
ISBN-13: 978-1505265736

For the Harmed

CONTENTS

ACKNOWLEDGMENTS	v
FOREWORD	1
INTRODUCTION	4

1 WHAT IS HYDRAULIC FRACTURING? — 6

Preparing the Site	6
Hydraulic Fracturing – the Detail	14
The Planning Process	15
Scottish Planning System	19
Northern Irish Planning System	20
Where Does the UK Stand at Present?	20

2 FRACKING: THE PROS AND CONS — 23

Social	23
Environmental	26
Economical	37
Political	44

3 UCG, CBM AND ACID WASH — 47

Underground Coal Gasification	47
Coal Bed Methane/Coal Seam Gas	50
Acid Wash	53

4 THE FRACK FREE STORY — 55

The First Frack Attack	55
Not for $hale	58
Looking South	61
Scotland's First Victory…	63
…And One Up for Wales!	63
The Fracking Threat Returns	64
Toxic September	65
All Quiet on the Fracking Front	66
Arise, Sir Frackman	66
The Big Rig Revolt	67
Lancashire – New Year, New Fracks?	69
More Eyes on Sussex	70
The Battle of Balcombe	72
England's Green and Pleasant…Water?	79
Double Trouble for Barton Moss	79
Winter 2013 – Spring 2014	85
Falkirk in the Spotlight	86
Secret Shale	87
2014: Five Blockades in Four Months	95
Back to Balcombe…	99
…Lancashire in the Limelight	102
Balcombe to Belcoo	106
Frack Attack – Round Two	108
Reclaim the Power	109

Rathlin Warned	114
Celtique's First Well	114
The People's Climate March	118
October 2014 – Two More Camps	119
Borras Borehole	121
IGas Eyes on Sussex	123
News from West Newton	124
A Centre for Fracking Excellence?	125
More Councils Turn Frack-Free	126
Walk the Walk '14	129
Autumn 2014: Three Warnings in One Month	130
Frack Attack – Round Three	132

5 OTHER KEY POINTS OF THE FRACK FREE MOVEMENT — 134

Protest Policing and Covert Intelligence	134
NATO Enters the Debate	142
Council Investments	143
European Union and Shale Gas	144
The Transatlantic Trade and Investment Partnership (TTIP)	154
The Infrastructure Bill 2014	155
Druids Against Fracking and The Warriors Call	156
Wrong Move Campaign	157
Talk Fracking	158

Observer Ethical Awards 2014	161
Ecotricity	161
APPENDIX ONE	163
APPENDIX TWO	173
APPENDIX THREE	187
GLOSSARY	216

ACKNOWLEDGEMENTS

Special thanks go to Graham Bentley, Jono and Paula for kindly donating their photos; to Helz for her advice throughout the production of this book and for bringing all the contributors together; to Lorraine and Jill for all their support and careful editing; and to James and Sue for their kind support, proof-reading and for writing the Foreword for this book.

I also wish to express deep gratitude to the many wonderful members of the Marcellus Outreach Butler group in Pennsylvania for compiling the list of fracking-related accidents, incidents and catastrophes that have occurred in the USA and Canada and for kindly allowing me to reproduce their work.

Finally I must pass on my appreciation to all the protectors, letter writers, petitioners, sign makers, artists, researchers, scientists, geologists, engineers, hydrologists, keyboard warriors and campaigners everywhere for their tireless efforts to bring fracking to public awareness and for their selfless and gallant acts of peaceful resistance.

PICTURE CREDITS

© P. Astridge: Fig.1; Fig.13; Fig.15; Fig.21; Fig.22; Fig.23; Fig.27; Fig.28; Fig.29; Fig.30; Fig.31; Fig.32; Fig.33; Fig.34

© G. Bentley: Cover; Fig.10; Fig.11

© M. Dale: Fig.4; Fig.5; Fig.6; Fig.7; Fig.8; Fig.9; Fig.25

© Jono: Fig.2; Fig.3; Fig.12; Fig.14; Fig.16; Fig.17; Fig.18; Fig.19; Fig.20; Fig.24; Fig.26; Fig.35; Fig.36; Fig.37; Fig.38; Fig.39; Fig.40

Acknowledgements

Other Sources Used in the Production of this Book

Britain and Ireland Frack Free, Concerned Communities of Falkirk, Extremeenergy.org, Falkirk Against Unconventional Gas, Fermanagh Fracking Awareness Network, Frack Free Arun, Frack Free Balcombe Residents Association, Frack Free Fernhurst, Frack Free Fylde, Frack Free Horsham District, Frack Free Network, Frack Free Ryedale, Frack Free Sussex, Frack Free Tyne and Wear, Frack Free World, Frack Free York, Frack Off Romania, Frack Off UK, Fracking Free Ireland, Keep Billingshurst Frack Free, Keep Kirdford and Wisborough Green, Keep Tap Water Safe, Residents Action on Fylde Fracking, Ribble Estuary Against Fracking, Schaliegasvrij.nl, Talk Fracking, Wirral Against Fracking, Worthing Against Fracking

BBC, Blackpool Gazette, Brighton Argos, Canterbury Times, CBC News, Chester Chronicle, Columbus Dispatch, Daventry Calling, Denton Record Chronicle, Gazette and Herald, Huffington Post, KJAS.com, KSAT.com, NBC4, New York Times, News North Wales, newsok.com, Post Gazette, Russia Today, Salford Star, Scandinavian Oil-Gas Magazine, Thanet Watch, The Advertiser, The Daily Mail, The Denver Channel, The Guardian, The Independent, The Mirror, The Observer, The Telegraph, West Coast Native News, West Sussex County Times, WBNG.com, WHEC-TV, WKBN.com, WPEC-TV, WTRF-TV, Yorkshire Post

Billingshurst Parish Council, Brighton and Hove City Council, British and Irish Legal Information Institute, Buckinghamshire County Council, Cambridgeshire County Council, Cumbria County Council, Department of Energy and Climate Change, Derby City Council, Derbyshire County Council, Directorate for Planning and Environmental Appeals, East Yorkshire County Council, Environment Agency, Fingal County Council, Gov.uk, Hansard, Lancashire Association of Local Council, Lancashire County Council, Newcastle City Council, North Yorkshire County Council, Planning Appeals Commission for Northern Ireland, Planning Inspectorate, Planning Portal, Pulborough Parish Council, Sheffield City Council, South Downs National Park Authority, West Chiltington Parish Council, West Sussex County Council, Wirral Borough Council, Wrexham County Borough Council

Campaign to Protect Rural England, Campaign to Protect Rural England Northamptonshire, Campaign to Protect Rural England Yorkshire and The Humber, Celtique Energie, Cuadrilla Resources, Drill or Drop, Friends of the Earth, Green Party of England and Wales, Greenpeace UK, IGas, Magellan Petroleum, Network for Police Monitoring, Open University, Plaid Cymru, Rathlin Energy, Tamboran Resources, Think Progress, Third Energy

Aswar.org.uk, bakken.com, Clean Water Action, Colorado Energy News, Eytanuliel.com, drillingahead.com, Food and Water Watch, Google Earth, Google Maps UK, Insdieclimatenews.org, Insidemedia.com, Marcellus Monitor, Natural Gas Watch, Oil-price.net, Online Conversion, pennenergy.com, Protecting Our Waters, Streetmap.co.uk, Tapnewswire.com, Thinkprogress.org, VoteWatch Europe, wave3.com, Wrexham.com, yourerie.com

FOREWORD

Anyone who's concerned about the future of this country should have a copy of this amazing book.

Make no mistake - fracking in the UK would affect all our lives, whether directly or indirectly. And not for the better!

Martin starts by explaining, simply and clearly, what fracking is. He lays out the facts about the mechanics of the industry, how it all works, and what's involved and there are a lot of things it would be wise for us to know!

He then goes on to provide an extensive list of the activities and campaigns of the growing number of UK groups who now question this invasive process. There are comprehensive lists and contacts for organisations across the country and further afield. His detailing of various incidents is always backed up by evidence from people involved, or by legal and medical records.

This book covers everything – political involvement – financial implications – environmental and health risks - scientific opinions, both pro and anti – security policies – UK action groups - overseas activities and evidence from the USA, Australia, Canada and others – industry myths - media coverage - in other words, pretty much all we need to know so far! Martin has painstakingly gathered together what we can only describe as an 'encyclopaedic reference library' on all aspects of this industry.

A bonus for us is Martin's experience as a Local Councillor. This means that he knows his way round bureaucratic institutions better then many of us, he speaks the language of District and County Councils and their various committees, understands the way they work, and he's not afraid to take on the corridors of power.

He also has an indomitable spirit and energy to seek out truths through Freedom of Information approaches in many different areas of governance. The results of this are here for you to read!

We are honoured to be asked to write this forward. Like so many people nowadays we have grave concerns about all fossil fuel industries, and about shale oil/gas exploitation here in the UK in particular. We worry about further congestion on our roads, about noise and disturbance, the disruption of communities, destruction of wildlife habitats, and possible long-term pollution implications.

And what of the future for this small island of ours? Nobody can answer that. There is a marked lack of balanced debate in the mainstream media. If the government did allow companies to put most of the contaminated water back into the boreholes, for so-called "storage", what long-term changes are likely deep underground? The numerous geological faults across the country mean aquifers are at risk, threatening our water supplies and serious environmental damage over time. The oil companies are forced to admit that - **all wells leak eventually** – even after the drilling is long over. But they'll have taken the money and run by then of course!

Almost two years ago now, Emeritus Professor of Geology David Smythe said something that we can't get out of our heads. He said **"chemical substances pumped under pressure into the ground will remain there for archaeological aeons "** We can't begin to imagine how long that is, but we don't want this kind of legacy for our children, our grandchildren and their children too. Nor does anyone we meet who has begun questioning the government and industry's famous **Dash for Gas.**

This book will really help you find some answers.
Thank you Martin.

James Bolam and Sue Jameson

<u>NB</u>– if you are a bit of a sceptic about all this – maybe have a look at appendix 3. There Martin has printed a diary from a group in the USA recording all the incidents of leakages, explosions, traffic accidents, flouting of regulations and injuries etc that occurred over about 22 months in 2013/14. A sobering read!!

INTRODUCTION

Hydraulic Fracturing hit the headlines in the summer of 2013 when a roadside verge half a mile from the small West Sussex village of Balcombe was broadcast on TV screens across the world. Since then it has become a common subject of debate in households and town halls all over the United Kingdom. What began as a campaign by a few dozen people soon expanded into a nationwide social movement involving tens of thousands. From aristocratic landowners to council house tenants and from the elderly to the very young, people of all backgrounds, races, religions and political persuasions have united in a way unseen for a generation.

At the time of writing, the Conservative-Liberal Democrat Coalition Government, with the support of the parliamentary Labour Party, are legislating for a complete reform of ancient land ownership rights that trace back as far as the signing of the Magna Carta in 1215AD. The proposals would allow for drilling and injection of "any substance" underneath any piece of land or property as shallow as just 300 metres (1,000 feet) without the need for permission, whilst the Department for Energy and Climate Change is expected to announce the winning applications for exploration and development license blocks covering some 60% of the country.

On the other hand a small, but growing, number of local authorities in England, Wales, Scotland and Ireland are defying Government guidelines and policy by declaring themselves Frack Free Zones. Landowners in at risk areas, such as the village of Fernhurst in West Sussex, have also been taking steps to form a legal blockade around proposed drilling sites in an attempt (whilst they still retain their underground access rights) to prevent shale oil and gas exploration from taking place.

This book aims to tell the story of the UK's anti-fracking movement from its early beginnings in Lancashire in 2011 through to the present day, and to give an insight into the arguments put forward by both sides of the debate. To achieve this, the first three chapters are designed to give a semi-technical overview of what hydraulic fracturing involves and how it differs from conventional oil and gas drilling, as well as an explanation of how the various planning systems function. We then move onto a rundown of the arguments that are most commonly used by both the pro- and anti-fracking advocates before taking a look at some of the other unconventional fossil fuel extraction methods that are being used, or otherwise proposed for use, in the UK.

In Chapter Four, we explore the Frack Free Movement in some detail and the final chapter is a look at some of the other factors at play in the story of unconventional oil and gas in the United Kingdom – both those factors that are supporting the roll out of hydraulic fracturing and those that are joining in the campaign to stop it. A small number of appendices have been included at the end of this book that provide information of interest in the wider perspective, such as a listing of all anti-fracking groups in the UK and Ireland, a list of global fracking bans and finally a partial list of confirmed incidents related to hydraulic fracturing in the USA and Canada between January 2013 and December 2014. Throughout the book a number of key terms can be found in bold font and so a glossary has also been included to provide an explanation as to what they mean in order to give you a better understanding of the subject.

In all, this book aims to provide a detailed overview of the state of unconventional oil and gas development in the UK and help you to understand what it involves, how it is being opposed and, ultimately, for you – the reader – to decide for yourself whether you support or oppose the exploitation of shale oil and gas.

1
WHAT IS HYDRAULIC FRACTURING?

Hydraulic Fracturing – or 'Fracking' – has been used by the USA in its rush for natural gas extraction to bolster its domestic supply over the past decade. It involves the injection of water, sand and a cocktail of chemicals under extremely high pressure through a **borehole** drilled into layers of **shale rock** deep beneath the Earth's surface in order to release small pockets of natural gas or oil that is trapped within it. So how exactly does this process work?

There are two main phases in any drilling operation – the planning phase and the operational phase. We will begin by exploring the operational phases of conventional and unconventional oil and gas drilling, in an ideal, textbook-style situation.

Preparing the Site
The first step in the drilling of a typical oil or gas well (conventional or otherwise) is to prepare the ground upon which the **Well Pad** is to be situated. An area of land averaging around four acres in area is cleared of any obstructions and an access road is laid to connect it with the existing road infrastructure. The top soil is cleared and stored, often in bunds around the outer perimeter of the site, the ground leveled and layers of impermeable membranes are

Fig.1: Site preparation at the Broadford Bridge-1 well site in West Sussex, September 2014

rolled out across the ground. On this is then placed a layer of compacted stone gravel which then forms the working surface of the well pad. The impermeable layers usually consist firstly of a clay barrier laid directly upon the bare ground followed by thin, water tight fabrics rolled out in strips on top of the clay, with each strip slightly overlapping the previous one to ensure that there are no gaps. A trench is also dug around the outer edge of the pad, similarly lined with impermeable layers, in order to trap any water or other runoff and prevent contamination of the surrounding land. Next, a security fence of three to four metres in height is erected around the pad for the protection of the on-site workers and to prevent unauthorised access to the equipment. Once the pad has been laid and prepared, the drilling equipment and associated infrastructure can be brought onto the site.

A number of portable cabins, along with power generators, are erected on the pad for use as worker accommodation, toilets, canteen and offices. Storage containers for the drilling fluid, rock cuttings and the oil or gas are installed, and water storage areas are created for use in the drilling and also as a standby in the event of fire A small area at the edge of the pad is set aside and prepared as a small car park so that the pad is entirely self-contained. Next, a three metre circular pit is dug at or near the centre of the well pad and a concrete chamber – known as a

Figs.2 and 3: Balcombe well pad

Cellar – is sunk into the ground, which provides the foundation for the drilling rig and a secure compartment to contain any spillages and leaks from the surface infrastructure. In the base of the cellar, a large diameter **Conductor Pipe** is inserted into the ground to a depth of around 65 feet. The **Workover Rig** can now be brought on site, and is assembled above the cellar.

With the rig assembled, a **Drill Bit** is mounted on the end of the **Drill Pipe** and the drilling can begin through the conductor pipe that has already been installed. As the drill cuts through the rock, a mix of water and additives known as **Drill Mud** are inserted into the well via the drill pipe and through nozzles on the drill bit in order to cool and lubricate the bit and flush rock cuttings to the surface so that they can be extracted and stored ready for removal from the site. The mud is also designed to line the walls of the borehole to keep it intact and stabilise it against the pressure applied from the surrounding geology. Since the first stage of drilling takes the borehole through near-surface **Aquifers**, the mud is typically freshwater based in order to avoid contaminating the ground water. The drilling continues until the borehole extends to just below the deepest freshwater aquifer that is nearest the surface and the drill pipe and bit are removed.

Surface Casing is inserted into the borehole in order to isolate the freshwater aquifer strata and to serve as a foundation on which to anchor the **Blowout Preventer** on the surface. Two or three further layers of steel casing and cement are inserted into the surface casing to act as a safeguard against contamination and also to further stabilise the well from external pressures. Once all the casing is set in place, the borehole is pressure tested to check for any leaks. In order to cement the casing into place, the liquid cement mixture is pumped into the casing under pressure and forced out of the **Casing Shoe**. The high pressure means that the cement is pushed up towards the surface through the **Annulus** which creates a further seal to isolate the borehole from the aquifer. This step is repeated for each layer of casing inserted into the well. Once set, the cementing process creates a solid plug inside the casing and so the drill bit and pipe are reinserted to drill through the plug and continue down into the rock below.

The depth of the drilling varies on a case by case basis, depending on the target reservoir layer, but the vertical section is continued down until a point, known as the **Kick Off Point**, around 500 feet above the planned horizontal leg so that the gentle curve can begin to allow the horizontal borehole to be drilled. Typically, the mud used in this part of the drilling process – that is below the freshwater strata – is usually brine or oil-based including, ironically, the use of diesel. Up to now the process has been identical as for the drilling of a conventional vertical well.

Once again, the drill bit and pipe are removed from the borehole and a Downhole Drilling Motor with Measurement While Drilling instruments is lowered into the well to begin the angled drilling of the curve, continuing until the borehole reaches an angle of 90 degrees to the vertical section. The typical distance between the kick off point and the start of the horizontal section is a fraction under a quarter of a mile in distance, allowing for a gentle, gradual curve to prevent blockages or snagging of equipment. The drilling of the horizontal leg – or **Lateral Well** – can now begin. The process used is exactly the same as for the vertical section. Additionally, at various stages throughout the drilling of the lateral leg, the drill bit and pipe are removed to allow for cutting tool and drill bit changes to take place in order to better suit the geology of the layer being drilled. This process is known as **Tripping Pipe**.

Once the target distance along the lateral is reached, the drill bit and pipe are removed for the final time and **Production Casing** is inserted along the full length of the well bore, and cemented into place using the exact same process as for the surface casing earlier in the process. The cementing permanently secures the casing into position and seals the borehole to prevent **Hydrocarbons** and other contaminants from leaking into the surrounding geology or up to the surface via the annulus. With the well now completed, the rig is no longer required and a temporary **Well Head** is installed so that the pad can be prepared for the service crew to move onto the site ready for the production of the oil or gas.

Since the casing and cementing forms a solid barrier between the well and the oil or gas reservoir, the casing needs to be perforated so that the hydrocarbons can enter the well and flow to the surface. To do this, a **Perforating Gun** or 'Perf Gun', is lowered into the well through the casing by a wire line to the designated target section of the lateral section. An electrical current is sent down the wire to the Perforating Gun and a series of explosive charges are detonated, creating small holes through the steel casing, cement layer and a short distance into the surrounding rock formation and the gun is removed from the well bore. Up to now, the entire process has been identical to the drilling of a conventional horizontal well, and this point marks the difference between where conventional ends and unconventional begins.

Shale rock is a **Tight Formation** unlike the traditional porous limestones targeted by conventional drilling. This means that the oil or gas is not free flowing and so the well needs to be stimulated by hydraulic fracturing. A detailed description of fracturing follows shortly, but as an overview the process in its simplest form forces water, sand and additives under extremely high pressure into the well and out through the perforations into the surrounding rock. The pressure applied is so high that the shale rock slits or fractures and so allows the oil or gas contained within the rock to flow into the well.

A temporary plug inserted just before where the perforations and the fracturing have taken place so as to isolate that section from the rest of the well until production is ready. This is known as the Heel. The Perforating Gun with a new set of charges is reinserted and lowered into the well to a point approximately 50 to 80 feet before the first perforated section and an electrical current sets of the charges. Once again, the gun is removed and a temporary plug inserted. This process is repeated along the full length of the lateral well. The fracturing is necessary when exploiting tight formations such as shale, since these formations do not contain naturally occurring fractures and so production of fossil fuels would not otherwise be economically viable. Once all the fracking stages have been completed, the temporary plugs are either removed mechanically, or other wise drilled out depending on the type and method of plugging used so that the oil or gas can flow up and out of the well bore. The temporary well head is removed and replaced with a permanent one called a **Christmas Tree** and other necessary surface equipment is brought on site and installed if it hasn't already. A pipeline from the pad is then built to transport the oil or gas and connect with the major transport network. As more and more pads are developed across the oil or gas bearing formations, additional pipelines are built.

What is Hydraulic Fracturing?

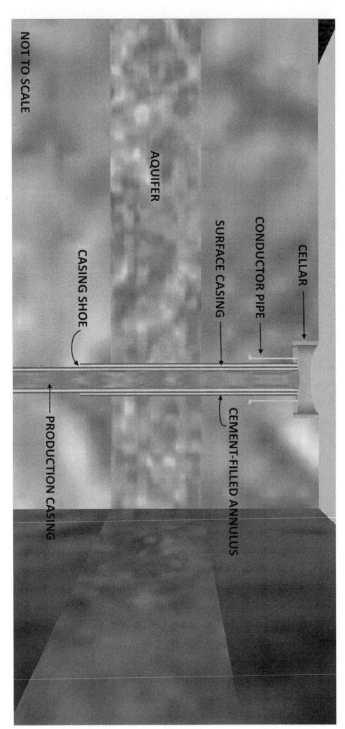

Fig.4: A typical well casing arrangement

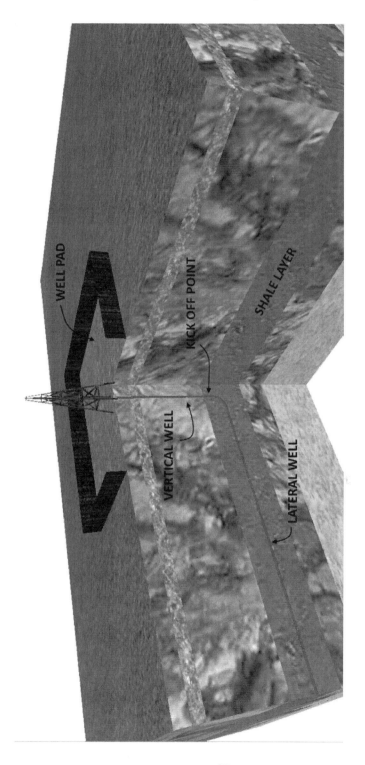

Fig.5: A representative illustration of an hydraulic fracturing well

Hydraulic Fracturing – the Detail

Returning now to the stage in our hypothetical well where the perforations have just occurred: - the workover rig that has been in place throughout the drilling is disassembled and removed and the specialist fracking equipment consisting of high pressure pumps and blending equipment moves in. This equipment will provide the means by which the water, sand or other **Propant** and additives are mixed in the correct quantities to form the **Fracking Fluid** and pumped at extremely high pressure down the well and into the shale formation. The first stage is for water from an external source to be pumped into on-site compartments, then into a hydration unit before finally ending up in a blender. The blender adds the propant and additives to the water and pumps it to the pump lorries – usually six or eight per pad – which insert the cocktail into the low-pressure end of the pumping manifold. Specialist hydraulic fracturing pumps within the manifold increases the internal pressure and forces the mixture out of the high-pressure end and into the well bore. The mix then begins to fill the well as more and more water is added until such point that the pressure reaches the extent where the shale fractures.

Fig.6: Fractures created in the shale rock through perforations in the well casing

The propant contained within the fracking fluid allows the fractures to remain open as the liquid content is drained from the well and so allows the oil or gas within the rock to connect with the well bore. A temporary plug is inserted at the Heel and the perforation and fracking process is repeated. In a typical operation between 8 and 12 stages of perforating and then fracking takes place per well. Once all the fracks are completed, the plugs are removed and the production of the oil or gas begins.

The **Produced Fluids** are diverted at the well head through a flowback manifold and into separate storage containers where they are then either recycled for use in the next fracking operation, or otherwise disposed of. If oil is the targeted product, then this is also diverted at the well head into separation tanks where naturally occurring gas can be removed and **Flared** off. Gas is always present in oil reservoirs – both conventional and unconventional – and is what provides the pressure to force the oil into the well bore. Oil is more economical and so rather than process and clean the gas, it is cheaper to burn it off as a waste product instead.

The Planning Process

The planning process is one of the key moments in any drilling application, because it gives local communities and authorities the chance to assess and decide upon the pros and cons of a particular application. It also gives national bodies the opportunity to similarly assess any application against national criteria and policy. Having an understanding of the process is vital for anyone interested in engaging with the subject of oil/gas and fracking.

In reality, the planning process for a particular site would have already been ongoing for 2 or 3 years before the local community gets to hear about it, whilst the regional planning can have been in process for many years.

In England and Wales, the first step is for a company or group of companies to apply to the Secretary of State for Energy and Climate Change for a **Petroleum Exploration and Development License** (PEDL). The winning bidder is then granted a license by the **Department of Energy and Climate Change** (DECC). Each PEDL is valid for a series of three sequential periods, called terms, during which the operator carries certain expectations. The idea of dividing the PEDL into terms is so that the Secretary of State has the power to assess the progress at the end of each term and, if it is believed that the operator is not effectively utilising the land, then can rescind the license and begin the process from scratch. The first term – known as the Initial Term – lasts for six years, during which the operator must actively explore for oil or gas deposits with in the license area in a pre-arranged programme set out during the bidding process. If the work programme has not been completed within the six-year term, the license automatically expires. At this point, the operator is also expected to relinquish up to 50% of the license block not intended for exploration so that it can be re-licensed. The Second Term lasts for a further five years to allow for detailed appraisal and development. Again, if the agreed work programme has not been satisfactorily completed, then the PEDL automatically expires. The Third Term lasts for 20 years and is when the actual production of oil and/or gas is expected. Depending on the rate of production, the Secretary of State has the power to extend the production into additional Terms.

With a PEDL having been granted to a particular operator, the first step is to identify specific sites suitable for exploratory drilling based upon a number of factors, not least landowner support, and drawing upon an evidence base such as **Seismic Surveys**. With one or more sites identified and agreed with landowners, pre-application discussions can then begin with the Minerals Planning Authority (MPA) – either the local

County Council or Unitary Authority – to identify preliminary thoughts as to site suitability, likelihood of securing planning permission and potential obstacles that need to be addressed in any application. Often, the local Parish or Town Council is approached for informal and preliminary discussions on the planned development and the operator may decide to arrange a community consultation event at which the public can view display panels on oil and gas drilling and ask questions to company representatives. Often, the attendees are asked to complete a survey that the operator uses as evidence in support of their application. This is typically the first time that a community is made aware that a particular site has been identified for drilling. For shale gas operations, applicants are also required to carry out an Environmental Risk Assessment at this stage. The Minerals Planning Authority also assess whether the operator is required to carry out an Environmental Impact Assessment. By this time, the operator would usually have also commissioned consultants to undertake a variety of basic surveys, such as on wildlife, heritage, noise and traffic.

There is usually a gap of a month or more as the operator pulls together and finalises its submission to the Mineral Planning Authority and an application is submitted. The MPA assesses the application and either chooses to validate it or notify the applicant that it is not valid and advises on what needs to be done for the application to be validated. Once declared valid, the MPA advertises and consults on the application and the Environmental Impact Assessment if one has been completed. The MPA is required as a minimum to write to those households within close proximity to the application site informing them of the application and any deadlines in which representations can be made, and to post a notice at the nearest publically accessible area bordering the site, such as a lamppost alongside a footpath or road. The minimum period of consultation lasts for 21 days from the date the application is made live on the local Planning Portal website, though it is not unusual for the letters and notice not to be posted until a week or more later, reducing the amount of time for responses to be made. Statutory bodies, including the Environment Agency, Natural England, English Heritage and relevant departments within the County Council or Unitary Authority, are similarly given notice and advised on the deadline to make representations. Once the 21-day period has concluded, the MPA assesses the application and continues to liaise with the applicants, identifying issues, recommending amendments and attempting to overcome obstacles to any potential objections – the MPA is required under the 2011 National Planning Policy Framework to work proactively with applicants in line with a 'presumption in favour of sustainable development'.

Once the MPA is ready to determine an application, it is brought before the authority's Planning Committee of councillors and a decision is made. In an ideal situation, this committee meeting should take place within 13 weeks from the date of validation. There are three potential outcomes, either approval, refusal or deferral to a future meeting. If permission is granted, the MPA can decide to impose planning conditions and the operator can apply to the Environment Agency for relevant environmental permits. The Applicant is then required to notify the Health and Safety Executive no less than 21 days in advance of activity starting on site and the British Geological Survey is also informed that work is due to begin imminently. The Environment Agency also has to be notified of the intent to begin drilling under the Water Resources Act 1991. The applicant now approaches the Department of Energy and Climate Change once again for final granting of consent to drill a well. The Applicant is required to begin operations within a specified timeframe set out by the MPA, otherwise if no activity has occurred, then the permission expires. They are also required to complete the agreed operations within another imposed timeframe, at the end of which the permission automatically expires.

The other outcome from a Planning Committee meeting is refusal and the applicant is formally notified of the reasons for refusal and of the right to appeal the decision. If the Applicant decides to appeal – they have a three month period in which to come to a decision – an appeal is made to the Planning Inspectorate and all third party representatives are notified by the MPA. The Applicant can choose which method of appeal is preferred – Written Procedure; Public Local Inquiry; or Formal Hearing. Depending on which is chosen dictates the timeframe in which a decision is made, but each method is preceded by a site visit by the appointed Inspector and the reading of all documents and representation made during the application process. The Inspector's decision is final except in exceptional circumstances when it can be proven that he/she failed to take into account all the evidence supplied. Third parties do not currently have any right of appeal in England or Wales.

Scottish Planning System

In Scotland, the planning system is almost identical to that of England and Wales, with the main difference coming at an appeal. Applicants still have the three month deadline to register an appeal, though there is no national Planning Inspectorate. Instead, the body responsible for re-assessing an application varies depending on who made the initial decision. If a local authority Planning Officer made a decision to refuse under delegated powers, then an applicant can appeal to a Local Review Body where a group of elected Councillors either upholds or overturns the decision. On the other hand, if Councillors had made the initial refusal, then the applicant appeals to Scottish Ministers.

Northern Irish Planning System
The Northern Irish system again follows a similar decision making process, but is significantly different from the rest of the United Kingdom. In the current system, local Councils do not have the power over planning matters, with all decisions instead being made by the Department of the Environment's Planning and Local Government Group, and the process of bidding for a PEDL block is in the hands of the Northern Irish Department of Enterprise, Trade and Investment. The Department of the Environment is still required to undertake the same stages in the planning process as that undertaken by Mineral Planning Authorities in England and Wales. As with Scotland, there is no Planning Inspectorate in Northern Ireland as it is known in England and Wales. Instead, appeals are made in the same way to the Planning Appeals Commission, which is an independent body separate from the Northern Irish Government.

The major difference, however, is that in Northern Ireland oil and gas drilling can come under a process known as 'Permitted Development'. This is an understanding enshrined in planning policy that certain types of development can take place without the need to apply for permission. In England and Wales this is typically restricted to only very minor developments such as small conservatories, with minerals extraction being considered a major development. It is this process of Permitted Development that drilling company Tamboran have relied upon to begin their exploratory operation at Belcoo, near Fermanagh.

Where Does the UK Stand at Present?
The new technology of High Pressure, High Volume Slick Water Hydraulic Fracturing made its debut in the UK at Preese Hall near Blackpool in Lancashire in 2011, and at the time of writing remains the first and only fracked well site in

the UK. This operation by drilling company Cuadrilla famously resulted in two minor earthquakes that damaged the well and reportedly damaged a number of properties in the vicinity, though Government and industry deny this. As a result of the earthquakes, the operation was suspended and the Government imposed a moratorium on the use of hydraulic fracturing between June 2011 and December 2012 while investigations into the safety of the process were undertaken. Concluding that the technology is safe, clearance was given for a nationwide roll out. Although the moratorium has been lifted for over two years, there have not been any other fracking operations as of the time of writing, although there have been a number of applications for exploratory drilling across the UK, many of which are seeking to drill into the shale layers.

The Coalition Government are overtly pro-shale and there have been a number of attempts to reform the planning and regulatory systems to allow the rapid roll out of hydraulic fracturing operations, such as tax incentives for drilling companies, and financial offers to local authorities who accept the process of fracking in their area. One such reform is of landowners' trespass rights, so that operators can be permitted to drill lateral wells under land without needing to gain the express permission of landowners. Whilst officially cited as a way to relieve the companies of the requirement to seek out all landowners concerned – often a lengthy and complicated process, particularly if not all landowners are in favour of the drilling – it is the belief of many in the anti-fracking movement that this is a move by the Government in response to the 'Wrong Move' campaign by Greenpeace and a successful implementation of a legal block at Fernhurst in West Sussex in which all the landowners surrounding a proposed exploratory site have written to the operator, Celtique Energie, to deny permission to drill under their land. This led to Celtique amending their plans so that only a vertical well is now proposed. A similar blockade has recently occurred at Celtique's other proposed site between the villages of Kirdford and Wisborough Green, also in West Sussex. The 'Wrong Move' campaign has, at the time of writing, attracted just under 58,000 landowners and homeowners across the UK to sign up to a nationwide legal block. A consultation by the Government on the proposed changes to the trespass laws was due to end in mid-August 2014, and will help to determine the extent of any reforms. Another proposed change is to the legal requirement for applicants to notify nearby residential properties when an application has been submitted, where only a single notice in a public place would be the minimum requirement.

Additionally, a 14th round of onshore licensing was launched on 28th July for applications open to bid on PEDL blocks across the UK until 28th October 2014. Over 60% of the country has been made available, with winning bidders expected to be announced by mid-2015. When announcing the opening of bidding, the new Energy Minister, Matthew Hancock MP, stated "unlocking shale gas in Britain has the potential to provide us with greater energy security, jobs and growth", clearly indicating the desire to encourage applications for the unconventional extraction of oil and gas.

In the following chapter we will explore a brief overview of the arguments used by both sides of the fracking debate, so that you, the reader, can come to your own conclusions about any potential future use of the controversial technology in the UK.

2
FRACKING: THE PROS AND CONS

The arguments used in the fracking debate can be split into four main areas:

Social – how the people of Britain would be directly affected or might benefit from fracking in their area;

Environmental – how fracking would impact upon or benefit the environment and climate change;

Economical – how fracking and its alternatives would affect or benefit economic growth; and

Political – how fracking would impact upon or benefit our lives through political issues.

Social

Pro	Con
- Shale gas is a bridging fuel to keep the lights on as we make the transition to low-carbon energy	- Renewable forms of energy are much more efficient at producing the energy the UK needs to meet demand. The technology already in place.
- The amount of shale oil and gas in the UK could provide domestic supplies for decades.	- A recent 2014 report into shale oil and gas by the British Geological Survey highlights the fact that even if higher-end production rates occur (up to 10% oil/gas recovery rate), would only provide 3 to 4 months of UK gas demand at current rates, negating the argument

	that this is an effective bridging fuel.
- Fracking would provide up to 74,000 more jobs in the UK	- The study often cited about job creation was commissioned by the fracking industry, whereas independent studies have estimated that renewable energy can create up to one million new permanent jobs. - Any jobs created by fracking would only be temporary, since shale oil and gas, like all fossil fuels, is a limited resource.
- Fracking is a technology that has been used for over 60 years in the UK	- Artificially stimulating wells is an old technology, but high-pressure high-volume slack water hydraulic fracturing has only been in use for around ten years. - The Department of Energy and Climate Change, and the former Secretary of State for Energy and Climate Change confirm that only one well has been 'fracked' in the UK – Preese Hall near Blackpool in 2011.
- Fracking does not cause any harm, or pose a threat to health and wellbeing.	- An increasing number of peer-reviewed studies in the USA are finding links

	between proximity to fracking sites and increases in health issues (silicosis, birth defects, hormones disrupted, cancers, still births, breathing problems). - Similar problems have been experienced in regions of Australia where fracking is in use. - In April 2014, a Texas Court awarded a family living in close proximity to a fracking site $275,000 for reduction in the price of their property, $2.4 million in compensation for physical and mental harm and a further $250,000 for future harm. This ruling was upheld by a Judge in July.
	- Further studies have shown a dramatic increase in crime rate in towns neighbouring fracking sites in the USA, as transitory forms of crime (drugs, prostitution) follow the shale gas workers from town to town.
	- Fracking requires vast quantities of fresh water, as does conventional

	drilling. This would place extra strain on water-stressed regions of the UK such as the South East of England (where the Weald Basin is a target for shale exploration.
	- There would be competition for water for domestic use and the fracking industry, as is being seen in the USA.

Environmental

Pro	Con
- Fracking produces lower Carbon Dioxide (CO_2) emissions than conventional oil and gas wells and from coal.	- Although shale oil and gas produces lower levels of CO_2 than other fossil fuels, it actually produces large quantities of methane, which has been found to be 75 times more potent a greenhouse gas than carbon over 25 years.
- Evidence from the USA shows that significantly higher levels of climate changing greenhouse gasses are emitted from fracking operations than previously estimated, jeopardising the Kyoto Protocol and climate change targets.

- The low CO_2 emissions make fracking a suitable transition technology to low-carbon and renewable energy in the UK.	- The technology for efficient no-carbon renewable energy (including 'green gas') production is already in place, avoiding the 'need' for fracking and further fossil fuel usage. - Additionally, the chemical composure of the natural gas produced by fracking has been found to vary greatly depending on its depth (i.e. it's 'maturity') and geographical location. A number of incidents such as blowouts and pipeline explosions have been the result of unexpected high-volatility of the gas produced.
- The amount of water used in fracking is considerably less than used in agriculture and even by golf courses annually.	- The water used in agriculture and golf courses is able to return to the natural water cycle and so is not wasted or removed. Fracking, however, pollutes the water used to such an extent that it cannot be cleaned or returned to the water cycle. - The amount of water required for each <u>individual fracking well</u> is vast – comparing it to

	<ul><li>annual agricultural and commercial uses is nonsensical.</li><li>In the USA, the demand for water for use in fracking operations has exaggerated the water stress in dry regions, leading to widespread and severe droughts, and competition between fracking companies and communities.</li></ul>
<ul><li>Freshwater aquifers (i.e. drinking water supplies) are protected from any potential contamination by hundreds of metres of impermeable rock and by several layers of steel and cement well casing.</li></ul>	<ul><li>No matter how strong regulations are, equipment failure and human error can always occur. An example from the UK is the conventional oil well at Singleton in West Sussex where two leaks remained undiscovered for several years and took another 5 years to resolve. Furthermore, well casing is not of a uniform thickness throughout the length of the well, with weaker areas where leaks can occur.</li><li>A borehole is restricted in size, so there is a limit on how much casing can be inserted. The more layers of steel and cement</li></ul>

	inserted, the thinner each layer has to be. Over time, steel corrodes and cement deteriorates, breaking down the barrier separating pollutants from surrounding geology and freshwater aquifers.
	- Industry figures for worldwide onshore wells (conventional and unconventional) show that up to 5% of all wells drilled fail immediately, rising to 50% after 20 to 30 years. Eventually 100% of wells fail.
- Supposed groundwater contamination is by naturally migrating methane.	- Although natural methane migration can and does occur, there have been an increasing number of studies from the USA and Australia proving links between aquifer and groundwater contamination as a direct result of fracking operations. In the USA, four states have proven that their water sources have been polluted by fracking. The geology of the UK is much older and more heavily faulted than in the USA, meaning that there are many times more

	opportunities and likelihood of migrating contaminants and seismic events.
	- Additionally, the UK is regularly subjected to widespread flooding events. In Colorado in 2013, dozens of fracking pads were flooded and millions of gallons of contaminants entered the water, polluting a vast area of land.
	- If drinking water is safe, why are drilling companies in the USA going to the huge expense of shipping and supplying bottled water free of charge to residences near fracking sites?
	- Since fracking is a new and untested technology in the UK (with just one fracked well ending in failure), the UK has no regulations on fracking. In fact, all onshore drilling regulations are based upon American best practice guidelines.
-The UK has a Gold Standard regulatory regime for hydraulic fracturing.	- The promise of Gold Standard Regulations is a fallacy. Leaked documents from the European

Commission in January 2014 identified Prime Minister David Cameron and the UK Government as the chief opposition to new environmental protection legislation on fracking operations, stating in a letter to the EC President **"It is essential the EU minimise the regulatory burdens and costs on industry…by not creating uncertainty or introducing new legislation. The industry in the UK had told us that new EU legislation would delay imminent investment."**
- The leaked documents also indicate that the pro-shale lobby had a "recently agreed core script" that was being pressed by Ministers and other UK Officials to "see off" the new proposed legislation.
- Boreholes are actually a loose helix in shape as opposed to straight lines. This means that the cement casing varies in thickness throughout the length of the borehole.

- Only water is used as a lubricant and coolant when drilling through water-producing layers.	- Once drilling through the water producing strata, an oil-based lubricant and coolant is usually used since it is more effective. Sometimes this includes the use of diesel. - This oil-based lubricant and coolant poses an environmental pollution hazard through the processes already explained above.
- Additives (chemicals) only make up a very constituent part of drilling fluid - The chemicals used are harmless substances that can be found around the home.	- Although chemicals and additives make up only a small percentage (less than 1%) of drilling fluid, the quantities used are still huge and pose a significant threat to health and the environment if spilled or leaked. - Despite pro-shale lobbyist claims that all additives are harmless, Freedom of Information Act requests (FOIs) prove that hazardous substances are used, even in conventional and exploratory drilling, such as at Balcombe and Billingshurst in West Sussex. - Whilst the majority of additives can be harmless,

	it is also the naturally occurring toxic and radioactive materials (**NORMs**) that are brought up from deep underground in the **Produced Fluids** that pose a threat to health and the environment.
- 'Minor Seismic Event' is a better way to describe the 'earthquakes' experienced at some fracking operations. - The seismic events cause no damage and are barely detectable - Professor David Rothery of the Open University states in an online course that artificial, fracking-induced (or 'premature') earthquakes are beneficial in preventing larger earthquakes in the future	- The main issue regarding fault lines is of migration of contaminants into the surrounding geology and vertically towards the surface. - Fracking operations and the injection of waste water into non-production wells can induce seismic events, such as the 2.3 magnitude earthquake at the only UK fracked well in Lancashire in April 2011. - In the USA, there have been a large increase in seismic events experienced in fracking areas. For example, Oklahoma now has more earthquakes than the seismically active state of California. Within the first two months of 2014, Oklahoma experienced 103 earthquakes of

	magnitude 2.5 and greater, with nine occurring in just one day alone. In previous years, 3.0+ magnitude earthquakes increased in number from 277 in 2000-2013 to 375 in 2014 alone.
	- In several other States where fracking operations are increasing, a rise in earthquakes is also being observed, with an increasing frequency of magnitude 3 and greater earthquakes. This has led to scientists at the USA Geological Survey expressing caution at this pattern and warning that a massive tremor could hit Oklahoma in particular "at any moment".
	- In early 2014, the USA also experienced a large landslide believed caused as a result of frequent seismic events from local fracking waste injection sites.
	- A number of householders near the Preese Hall site reported damage to property from the two induced earthquakes in April 2011. Additionally, investigations discovered

	that the well bore had also been damaged along an almost 200 foot length by the earthquakes. A Government report found that injection of fluids into the well was responsible for the tremors. - The American Army Corps of Engineers responsible for flood defences will not permit any hydraulic fracturing within 0.9km (3,000 feet) of their infrastructure.
	- Aside from all the other environmental issues, if the UK is to see the same kind of 'fracking boom' as in the USA then large tracts of the countryside would have to given over to well pads and pipelines – approximately one pad every kilometre, each with up to 10 or 12 lateral wells.
	- There are an increasing number of reports of the effect of fracking operations upon wildlife populations from around the world, such as bird deaths from flying over gas fields and of mass deaths from leaks of

	contaminants into water courses. In the UK, residents at Balcombe in West Sussex have reported the absence of birds and other wildlife from gardens during and after the exploratory drilling in 2013.
	- Fracking creates vast quantities of contaminated water and fluids that require disposal. In the USA, this is often achieved via two methods – either the injection of the waste into other wells or long-term surface storage in containers, both of which bring about their own hazards. - Additionally, there are a number of cases of illegal and accidental disposal of toxic and radioactive waste in lakes and rivers, as well as by tankers spraying the roads. - In 2011, Cuadrilla was given permission by the Environment Agency to dispose of huge quantities of radioactive and contaminated waste water into the Manchester Ship Canal

	- UK pro-shale advocates cite that there are 'Gold Standard Regulations' that make the process of fracking safe for the UK. However, It has been found through Freedom of Information requests that the Environment Agency nor the Health and Safety Executive conduct independent inspections of any oil or gas well sites, but instead rely on self-regulation by the well operators.

Economical

Pro	Con
- Fracking will lower energy bills as experienced in the USA.	- The UK is tied into pan-European markets that regulate energy costs, so will not see the same kind of reduction in bills as the USA initially experienced. - Although the USA saw a substantial reduction in gas bills, their electricity and water bills saw a significant increase meaning that overall energy costs were not benefited by lower gas prices. - In the USA, the cheap gas

	prices lead to shale operators exporting gas to more expensive international buyers, mainly in Europe, to increase revenue and stay in profit. This negates any argument that shale oil and gas is an assured domestic energy source.
- Fracking will boost the national economy through revenue and taxation.	- The large number of HGVs required to service each well pad cause a significant amount of damage to roads and transport infrastructure that needs to be repaired at taxpayers' expense. - Examples of the repair deficit of fracking from the USA include Texas where in 2012, the State collected $3.6 billion in revenue from drilling, but estimated the cost of repair to roads as a direct result of drilling as $4 billion. Additionally, since 2009, Arkansas has taken $182 million in revenue but faces a bill of $450 million in road repairs from drilling operations. - The wear and tear on roads from fracking is disproportionate to

regular use from everyday traffic. For example, in 2012 the state of Texas produced a report investigating the impact of fracking traffic and found that on average it takes 1,184 loaded HGVs just to bring one gas well into production; 353 more HGVs per year to service each well; and 997 HGVs every five years to re-frack each well. The damage to the roads was estimated to be equivalent to 8 million cars per well and another 2 million cars a year for maintenance. Overall, roads designed to last 20 years were only lasting a fifth of their expected lifetime.
- The impact upon roads in the UK would be much more acute since many rural roads are narrow and not designed for use by HGVs.
- Aside from the award of $275,000 to an American family in April 2014 for reduction in property value, similar impacts upon house prices have already been felt in the UK

although there has been no fracking since 2011. Estate Agents and sellers are reporting reduced prices in areas close to exploratory drilling and several insurance companies are refusing to offer cover for properties in identified shale areas. In 2014 one property near Blackpool was reported to have fallen in value from £725,000 to less than £200,000 following new fracking applications nearby.
- A Government report in August 2014 on the affects of fracking on rural communities identifies a list of impacts upon many social and economic factors, including property prices, but have redacted almost the entire report.
- Above all, fracking is not a cost-efficient technology. Production peaks after a year and each well requires re-fracking after 2 to 5 years, with most wells expiring after just three years, necessitating the need for more and more new wells to be drilled

- every year. The expense of this short-term production means that in the USA, operators are spending out between $1.50 and $5.00 for every $1.00 they make.
- The fracking industry is heavily reliant on low-interest loans and investments to fund the programme of drilling – 2,500 wells per year for the Baaken shale prospect in the USA – so that production can meet up with demand. Financial analysts have labelled fracking as a 'Ponzi Scheme' and reported that if investors refused to keep paying in money, or the Government raises interest levels, the entire USA shale programme would collapse, causing a recession deeper than the one of 2008.
- Furthermore, everywhere that shale oil and gas has been exploited or explored, reserves have been grossly overestimated – which some economists argue is a ploy to entice more

	investors. For example, in early 2014, the main shale prospect in the USA – the Monterey Shale, supposed to hold two thirds of the entire American shale gas – had a 96% write-down of its production potential. - In the UK, the Chancellor of the Exchequer, George Osborne MP, announced in 2014 that the Government will give large tax breaks to companies if they undertake hydraulic fracturing of shale reserves, further negating the argument that the nation economy would see a boom.
- Local communities who accept fracking will benefit financially through £100,000 grants, retention of business rates and 1% of profits.	- The financial incentives are little more than a 'bribe' by Government to entice communities to accept fracking. The costs associated with the damage done by fracking far outweigh any income through revenue. - Furthermore, the one-off community payouts are only voluntary, and drilling operators are not liable to make any payouts.

- Any money handed over to communities would not go into the pockets of those living near to the fracking sites, but rather split between local government bodies, such as the Parish/Town Council, District/Borough Council, County Council, Police and Fire Services, etc. It would be up to them to see how, when and where the money gets spent, meaning that it could go towards projects that would not be of much, if any, benefit to those most impacted by fracking, or even away from the area where the fracking has taken place.
- Community-owned renewable energy provides a much more cost-effective, cleaner, greener and longer-term alternative.
- In Germany, communities have become completely energy self-sufficient through energy co-operatives. The village of Wildpoldsried, for example, produces renewable energy to such

	an extent that it feeds 321% into the grid, earning the villagers €4 million annually (as opposed to the £100,000 one-off payment from fracking). Similar situations are repeated across Germany.

Political

Pro	Con
- Fracking in the UK would lead to less reliance upon foreign sources of fuels and in particular would give the UK independence from Russian energy imports.	- The UK does not import any oil or gas from Russia, and is instead mostly sourced from domestic supplies already. - Any foreign energy imports are sourced from stable and reliable nations such as Norway, Denmark and Qatar.
	- The fracking industry is too heavily involved in the UK Government and regulatory bodies. For example, the chairman of Cuadrilla, Lord Browne, was the lead non-executive in the Cabinet Office, advising on energy policy. He was appointed to the role by the MP for Balcombe, Francis Maude. - The newly appointed

	Chairman of the Environment Agency, Sir Philip Dilley, was chairman of consultancy company Arup Group from 2009 to 2014. Arup was the Agent for Cuadrilla's drilling application to West Sussex County Council on 22nd January 2014 for a return to Balcombe. Arup is currently acting as the Agent for Cuadrilla's hydraulic fracturing applications at Preston New Road and Roseacre in Lancashire.
	- Of all five of the main UK political parties, only the Green Party opposes fracking.

Fig.7: Some of the dangers and risks that have been associated with hydraulic fracturing and unconventional oil and gas extraction

3
UCG, CBM AND ACID WASH

Hydraulic Fracturing is not the only unconventional form of extreme energy extraction that is occurring or planned for the UK. Here, we take a brief look at the technologies that are being termed "Fracking's Evil Twins". It is important to understand that there are any number of variations as to how a well or wells can be drilled, such as multiple wells and varying borehole trajectories. The descriptions and illustrations that follow are purely illustrative of some of the basic setups that have been used, or are proposed for use in the UK.

Underground Coal Gasification

Underground Coal Gasification (UCG) is a method to extract synthetic gas from coal seams that are too deep for conventional mining operations, by combusting the coal in situ underground and extracting the gases via wells drilled into the strata. The process has been trialled intermittently since the 1930s in the former Soviet Union, Europe and the USA (although the theory of gasification of coal was first made in 1868) and more recently in New Zealand, China and Australia.

The planning and site preparation processes are the same as those previously detailed for hydraulic fracturing. Once the pad is ready, two or more vertical wells are drilled into the target coal seam. One of the wells will become the injection well, where oxygen and/or steam is injected into the oxygen-deficient coal and where the coal is ignited. Once alight, the oxygen or steam maintains the combustion of the coal seam at around 1,000 to 1,200°C to firstly artificially fuse carbon atoms from the coal with oxygen to create carbon dioxide, and then initiates secondary reactions leading to the creation of carbon monoxide, hydrogen and methane, producing a synthetic gas, known as **Syngas**. Secondly, the injection well creates the pressure required to force the gas towards the second vertical well – the production well – where it can be transported to the surface for collection and cleaning before moving on to energy suppliers.

A variation on the process is that a horizontal well is drilled from the ends of the two vertical sections through the target coal seam, with a third vertical well drilled from a second pad to connect with the ends of the laterals. The two horizontal wells retain their functions as injection and production wells, whilst the vertical well acts as the point of ignition for the combustion to begin. This method allows for a strip of the coal seam running between the two horizontal wells to be combusted in a controlled path, as well as allowing for identical 3-well configurations to be drilled in series along the targeted coal seam as a form of underground strip mining.

Anti-fracking activists are equally opposed to this method of fossil fuel extraction on the ground of huge carbon emissions, subsidence, large scale groundwater pollution and the risks associated with the highly toxic and carcinogenic waste material left behind by the partial combustion of the coal and from the above-ground cleaning process. They argue that most tests of the process by the Soviets and Americans resulted in some degree of water contamination via leaks, compromised well integrity and natural fractures in the geology leading to migration of fluids as a result of the change in pressure in the coal seam. UCG has not gone into commercial production since the process has failed to meet environmental standards. Two well sites in the former Soviet Union did supply Syngas for a while, although only on a very small scale to power local factories until large reserves of cheaper natural gas were discovered. In April 2014, the

Australian Government filed a AU$2 million lawsuit against Linc Energy over alleged serious environmental damage of land and water from a trial UCG operation. A similar situation occurred with Cougar Energy in 2010 after an incident allegedly led to groundwater contamination of benzene and toluene, which were also detected in nearby livestock.

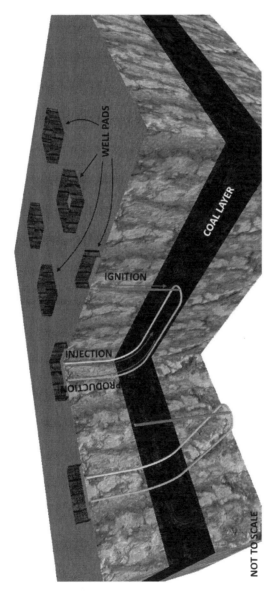

Fig.8: A representative illustration of a typical Underground Coal Gasification operation

Coal Bed Methane/Coal Seam Gas

As with Underground Coal Gasification, Coal Bed Methane (also known as Coal Seam Gas and Coal Mine Methane) targets deep coal seams that are uneconomical to be mined using conventional methods. However, unlike UCG, Coal Bed Methane is designed to extract gas held naturally within pores and fractures within the coal itself. CBM is also a much more recent, yet more advanced, technology developed in the 1970s in the USA as a way to collect natural gas from coal seams that would otherwise be wasted via venting bores drilled into the coal during conventional mining. It was developed into a stand alone extraction method in the 1980s. So far, the USA, Canada and Australia are the only countries to have moved into commercial CBM production.

In the USA, the first CBM well was drilled in 1980 in Alabama, with rapid deployment of the technology across Alabama, Wyoming, Colorado and New Mexico in the ensuing decades, reaching a peak in production in 2007. It is said that the USA has drilled over 55,000 CBM wells in the past 10 years alone. In Canada, Alberta is the only province to have entered commercial CBM production, with over 8,000 wells. Production in Australia began in Queensland in 1996 and has since spread into neighbouring New South Wales.

In the UK, methane from coal had only previously been extracted via ventilation of coal mines, but in 2008, a number of CBM exploration licenses were issued. In 2012, IGas began the first, and so far only, commercial production of CBM in the UK, though on a small scale.

The methods for CBM drilling are perhaps the most varied of all the technologies discussed in this book. First, we will take a look at how the IGas wells in the UK have been drilled. The site preparation and planning processes are, once again, the same as explored in the chapter on hydraulic fracturing, except that in the case of CBM, two well pads are constructed approximately 400m apart. On the first pad, a vertical borehole is drilled down to approximately 1,000m – some 80 to 100m below the lowest target coal seam. Once this drilling has been completed, a different drilling rig arrives at the second well pad site in order to drill a horizontal well. The first vertical section of this second well is drilled down to the top of the targeted coal seam before the curve building process begins for the horizontal section, which connects with the first vertical well, and then continues on for approximately another 1,000m. Further horizontal sections are drilled at deeper depths through all the targeted coal seams from the second well site, all of which intersect the first vertical well. Production is now ready to begin. The lateral wells become the injection wells, where water and additives are inserted, whilst the single vertical well becomes the production well, where all water and gas is collected. With the well full of water, the first stage is to drain it of the waste water and so beginning the release of gas from the coal. As the water levels, and therefore pressure, decreases, the natural pressures of the gas within the coal forces it out and into the well bore. The more water removed from the well, the more gas is released.

Once the gas begins to flow, testing is carried out at the vertical well site, including flow rate, flaring and chemical composition of the gas. If production is favourable for economic recovery, then the vertical well pad is reduced in scale to approximately one third of its original size and the horizontal well pad is reduced to around one twentieth, so that it comprises just the well head and a protective cage. At this point, the second pad is redundant, since it is non-producing, though some CBM wells require hydraulic fracturing in the future, as gas production declines. If, however, flow rates prove non-favourable, then either the wells are plugged and abandoned, or hydraulically fractured to stimulate the wells, and the flow testing repeated.

A variation on the drilling method is to utilise just a single vertical well that is drilled to just below the lowest targeted coal seam, perforated and hydraulically fractured to allow faster draining of the waste water into the surrounding geology and to improve the gas flow into the single well bore. A final variation being developed in the USA is to combine the two previous methods so that one vertical well is drilled, with lateral legs stemming off and into all various layers of targeted coal seams.

As with UCG and Fracking, CBM has its fair share of criticism. Although the combustion of the methane has a lower greenhouse gas effect than the straight burning of coal, methane is 75 times more potent a climate changing gas than carbon dioxide over 25 years. Furthermore, CBM wells have persistently resulted in the leak of methane into the atmosphere. In the USA alone, these **fugitive methane** leaks are responsible for 10% of all annual methane emissions. CBM, like fracking, creates vast quantities of produce fluids that can contain a cocktail of naturally occurring chemicals, salts, radioactive materials and heavy metals. In the USA in 2008, 1.23 billion barrels of produced fluids were collected from CBM activity alone, of which some 30% was disposed of into surface watercourses. Of the remaining 70% that is too polluted for surface disposal, much ends up being re-injected into wells. Furthermore, CBM wells are typically short lived like shale gas wells, and require more and more wells to be drilled, and potentially fracked. Other concerns arise over matters such as the extremely large quantities of water required which can deplete aquifer reservoirs. In Australia alone, the National Water Commission has estimated freshwater extraction for CBM use as over 300,000 million litres annually. CBM attracts much of the same arguments as for shale oil and gas.

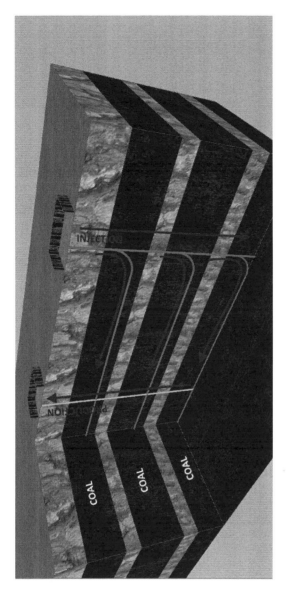

Fig.9: A typical representation of a Coal Bed Methane/Coal Seam Gas well

Acid Wash

Acid washing (sometimes known as Acid Fracking or Acidification) is a stimulation method that can be used in conventional and unconventional wells of all types. It is not a drilling method in its own right, but rather a way to improve the permeability of the targeted geology to improve connectivity of the oil or gas reservoirs with the well bores and to increase the flow rate. Acid Washing inserts large quantities of acid – typically hydrochloric acid – under pressure into the well bore and out into the surrounding geology through the perforations. The pressure is high enough to force the acid into natural fractures in the rock, but is low enough so as not to artificially fracture rocks as in hydraulic fracturing, though acid washing can be used in association with fracturing in shale oil and gas wells. The acid dissolves minerals such as clay and quartz that can naturally fill fractures in the target reservoir and inhibit permeability and production. This method has been proposed most famously at the Cuadrilla well in Balcombe, West Sussex, after the company announced that it would not frack the well, but would instead inject acid under high pressure to dissolve the rock instead.

Concern regarding the use of this method of stimulation typically encompasses the wider issue of fossil fuel extraction and greenhouse gas emissions, but also the contamination risks of injecting large quantities of hydrochloric acid through freshwater aquifers and into the rocks where natural faults could act as a conduit for migration. Other concerns include the traffic generated for the transportation of the acid by road, risks of spillages and accidents and the fact that the acid can deteriorate the steel and cement well casing over time, leading to leakages of other contaminants.

4
THE FRACK FREE STORY

"Any kind of fossil fuel extraction is contributing to climate change. There are no clean fossil fuels. None of them are renewable, all are dirty. We already have the renewable alternatives and we must demand maximum investment in them immediately. In the light of the recent IPCC report and UN statement on climate change, for the government to encourage any kind of ongoing dirty fuel extraction and development, is criminally negligent. If fossil fuel extraction is criminally negligent then it is the duty of every responsible citizen to oppose this crime. We must stop the drilling using whatever peaceful action it takes to do so."
- *Peter Whittick, Anti-Fracking Activist at Horse Hill, Surrey*
8th November 2014

The Anti-Fracking Movement in the United Kingdom has been one of the most incredible and rapidly growing community-led movements since the Campaign for Nuclear Disarmament in the 1980s, and the suffrage movement of the late 19th and early 20th centuries. Encompassing all religious, political and societal beliefs and classes, people of all ages and walks of life have united in a spectacular mass movement, the likes of which have not been seen for decades. Beginning with small, regional grassroots awareness campaigns, the movement has grown into a national campaigning body within a matter of three years, garnering the support of a number of familiar, household names from music, fashion, film and TV. The oil and gas industry have recognised the impact that campaigners have had on the implementation of unconventional oil and gas extraction in the UK.

The First Frack Attack
In 2007, the Department for Energy and Climate Change granted the drilling company Cuadrilla a license for

exploration of unconventional gas on the Fylde Peninsula in west Lancashire. A site at Preese Hall Farm, near the village of Weeton, was chosen – less than 4.5 miles (6.9km) east of Blackpool's famous Tower. Being agricultural land, the first application by Cuadrilla to Lancashire County Council was submitted on 31st July 2009 to seek permission to change the use of part of the farm into a drilling pad and upgrading the access road for the drilling of an exploratory borehole and testing for hydrocarbons present in the shale rock layers. At this stage there had been no mention of hydraulic fracturing and permission was subsequently granted, with no objections to the application, on 2nd November 2009. Another similar application was submitted on 3rd December 2009 for a second exploratory site at Hale Hall Farm near the hamlet of Wharles, only 4.9 miles (7.9km) from the Preese Hall site. Permission for this second operation came on 3rd March 2010. A third application came on 3rd February 2010 for another site less than a mile north-west of the village of Singleton (not to be confused with the West Sussex village of Singleton, also the site of an oil well), for the construction of a well pad (permitted 22 April 2010). Like the Preese Hall site, neither the Hale Hall Farm or Singleton applications made any mention of hydraulic fracturing, with only passing mention of testing for gas in the shale layer of the Bowland Basin. Instead, reference to possible need for "stimulation" by pumping water under pressure into natural fractures was the only hint that The Fylde could become the UK's first fracking development. The latter application was the only one of the three to receive any objections, citing that the site is unsuitable for a drilling operation. On 26th July 2010, a fourth site was applied for at March Nurseries near the village of Banks, around 8.7 miles (14km) south of the Preese Hall site. Once again, there was no direct mention of hydraulic fracturing and no objections were logged. Permission was granted on 21st October 2010. A fifth and final application was made for a site 1.9 miles (3km) north-east of Lytham St. Annes (3.75 miles

[6km] south-west of Preese Hall) on 1st September 2010 and subsequently granted on 22nd November the same year. Only one objection was raised (citing too much traffic would be generated), to one support registered. Once again, there was no mention at this stage of hydraulic fracturing, with brief references only to "stimulation" by injecting water and sand into natural fractures. Within a year, Cuadrilla had chalked up five potential fracking sites with barely any resistance.

Fig.10: Cuadrilla's drill rig at Banks, Lancashire

The first site destined for drilling was the Preese Hall Farm pad, with works commencing on 16th November 2009 and the borehole was drilled between August and December 2010. All was fairly quiet until Cuadrilla made the decision to test-frack the well commencing 26th March 2011. During the testing, a 1.5 magnitude earthquake shook the land on 1st April, and another, larger earthquake measuring 2.3 on the Richter Scale struck on 27th May. An immediate halt to any further fracking activity was made and a Government-imposed moratorium on the process was put in place for surveys and reports to investigate the cause of these seismic events. Although Government and industry sources maintain that fracking is safe and that the earthquakes caused no damage, the 'Stop Fylde Fracking' websites details 49 separate complaints from nearby residents of damage to property that are alleged to have occurred during or immediately after the two tremors, including 34 complaints of damage and distress caused during Cuadrilla's seismic surveying alone. The complaints range from cracks appearing in plaster, a spate of collapsed water mains in the St. Annes area and cracks in road, patio and driveway surfaces, to disturbances caused by explosions and late night drilling, and troublesome telephone surveys on behalf of Cuadrilla. The UK had had its first taste of fracking, and it left a foul taste in the mouths of many!

Not for $hale

With one other well site drilled in the vicinity of Preese Hall (though with no hydraulic fracturing) and permissions for test drilling and fracturing of three more, the campaign to stop fracking in its tracks began to take off, with members of the local community coming together to express concern. By August 2011, enough people prepared to take action had united and on the 6th August, a small team scaled 500 feet (167m) up Blackpool Tower in the early morning to unfurl two giant banners to raise awareness of the fracking threat to the UK and to announce the launch of the new national coordination group for peaceful, non-violent direct action – 'Frack Off'. After the somewhat unwise use of high-pressure water by the authorities to soak those manning the banners, the protesters were forced to descend the tower and subsequently arrested on charges of aggravated trespass.

In the meantime, with operations suspended and having passed their deadline to abandon the site, Cuadrilla applied to Lancashire County Council for permission to extend its drilling operation until December 2013 on 14th June 2011. Contrary to the first application almost two years earlier, this one attracted 508 objections, citing the dangers of fracking and highlighting the fact that the two test fracks had both caused earthquakes. The emerging anti-fracking sentiment had well and truly engrained itself in the minds of the British public. Despite this, Lancashire County Council voted to grant Cuadrilla permission to extend their operations 18 months later on 23rd January 2013.

With the drill rig now sinking its teeth into the third test site near the village of Banks at the end of August, more and more people in Lancashire were waking up to what could happen on their doorstep. A series of public meetings, leaflet drops and door-to-door conversations cemented the feelings developing around the Fylde. The first anti-fracking camp – Camp Frack – was established near the village of Singleton to bring together campaigners from across the UK (by now, Brighton was becoming a hub for anti-frackers on the south coast) and ran between the 16th and 18th September 2011.

Fig.11: The Banks well site during drilling operations

Although this provided an invaluable opportunity for networking between campaigners, what became quite clear was that they were still in the minority. The majority of people around the Ribble Estuary were unconcerned about Cuadrilla's plans and seemingly happy for the fracking to take place. To most people in the vicinity, the drilling was just occurring harmlessly in a few isolated spots. Despite this, the Camp continued to make excursions to many of the nearby villages and communities to hold public meetings and hand out leaflets. On the final day of Camp Frack, a two-mile protest march to Cuadrilla's Singleton drilling site was planned – the first of its kind in the UK, but certainly not the last. With around 140 people taking part, the police were nervous and threatened arrest for anyone attempting to walk along the farm track leading past the drill site, although its use as of right by walkers had been well established. In the event, no arrests were made and the scene was set for more campaigning locally and nationally in the coming months and years.

With the report into the cause of the two earthquakes in April and May 2011 ongoing, Cuadrilla called a press conference at Blackpool's Imperial Hotel on 21st September to announce the results of its limited exploratory drilling. Frack Off in turn called for the first anti-fracking demonstration to take place simultaneously at the hotel, in anticipation that the conference would be told that the drilling had been successful and production in Lancashire was favourable. In fact, it was reported that the company announced reserves of 200 trillion cubic feet of gas, or some ten times the amount of gas under the North Sea. The more sceptical members found the timing of this astonishing announcement rather convenient since the Australian share markets had recently announced that Cuadrilla's major financial backer, A J Lucas, had been suspended from trading on the country's stock exchange. It also came a matter of days before a major industry conference on unconventional gas in London, at which Frack Off held another demonstration, handing out leaflets and engaging with conference attendees. A third noisy demonstration was organised outside the Department for Energy and Climate Change offices in London on 13th October 2011, where Cuadrilla were giving a presentation on their investigation into the two Lancashire earthquakes. The results of this report were made public by the British Geological Survey a few days later on the 17th, when it was announced that in all likelihood Cuadrilla had indeed been responsible for setting off the two seismic events earlier in the year through their hydraulic fracturing operation at Preese Hall.

Welcoming this announcement, which had effectively confirmed what local and national anti-fracking campaigners had been saying for the previous six months, Frack Off called on the growing movement to stage the first mass anti-fracking direct action demonstration in the UK to take place at a £1,500-per-ticket Shale Gas Environmental Summit held in London on 2nd November 2011. The day was to become the first national "day of action" against the industry. On the morning of the 2nd November, nine activists managed to enter Cuadrilla's drilling pad at Marsh Nurseries near the village of Banks and scaled the drilling rig and 'occupied' the site, forcing it to suspend operations for the duration. Although one security officer attempted to grapple one protester, the event was otherwise peaceful, with the site workers even bringing cups of tea to those attached to the rig. In the afternoon, a flash mob – or 'Frack Mob' – of 100 people dressed in gas masks and protective clothing to symbolise the health risks of fracking assembled outside the Shale Gas Summit. The aim was to block the road leading to the venue, disrupt the conference by making as much noise as possible and hold a peoples' assembly of fossil fuels and energy. A second occupation of the Banks drill site was made on 1st December when 8 cyclists from Bristol made a repeat of the November action, and scaled the rig in protest at fracking and to highlight the recent wave of permissions for exploratory shale and coal bed methane applications in the south-west of England. In the end, the drilling operations were halted for 13 hours with five of the protesters being arrested for aggravated trespass.

Looking South

At the opposite end of the country, Cuadrilla had submitted a planning application to West Sussex County Council on 25th January 2010 to "upgrade existing stoned platform and drill an exploratory borehole for oil and gas exploration" less than half a mile (0.8km) south-west of the small village of Balcombe, which sits above the Weald Basin – believed to be a major source of shale oil. As with the applications in Lancashire, there was no mention of hydraulic fracturing, with only short statements saying that the main aim of the Balcombe site was to test the shale layers and, with the same wording as used for the Preese Hall application, that "stimulation" might be required which would involve the pumping of water and sand under pressure into natural fractures in the rock. Despite being billed as a 'conventional' well site, even before fracking had a controversial name in the UK, it was evident that hydraulic fracturing was expected to take place (a letter from Cuadrilla to the Department for Energy and Climate Change released under a Freedom of Information request in 2014 proves that Cuadrilla did indeed intend to frack the site, claiming that Balcombe would not be economically viable without the use of hydraulic fracturing). With permission granted on 23rd April 2010, there had been no formal objections registered – echoing the situation in Lancashire that was in process at the same time. Cuadrilla began the pad construction on 28th September 2010 with the intention of beginning drilling in June 2013, once a rig became available.

Following the revelations from Preese Hall in early to mid-2011, the threat of a potential fracking site in Sussex resulted in the South Coast Climate Camp group organising a public meeting in Brighton on 15th December 2011 to discuss the fracking process and how the plans to drill at Balcombe could be resisted. Fifty people attended the meeting, marking a modest start to the Sussex anti-fracking community that went on to receive worldwide media attention in less than two years time. A follow up meeting was arranged for 4th January 2010.

The first public meeting in Balcombe took place on 11th January 2012, where 300 people – over 16% of the village population – packed into the Victory Hall where the CEO of Cuadrilla explained the proposals for the site (known as Lower Stumble). He was not well received, nor were representatives from West Sussex County Council and Balcombe Parish Council, with an unusually emotive audience, the likes of which had not been seen in Sussex for some time.

Back in Brighton, 15 members of Frack Off spent three weeks in January 2012 designing and painting a huge banner warning of the dangers of hydraulic fracturing, which was eventually hung outside Brighton train station, along the main London railway line in March, where it remained for four months. On the 27th March, East Sussex also became the first county in England where the Council passed a motion to express concern at fracking, and request that the members receive full details on the process and its dangers before they would entertain any applications submitted in the future. The motion was adopted unanimously across all political parties, despite the Conservative members reportedly having received a phone call from the then Energy Minister Charles Hendry MP urging them to reject it. However, the Council rejected a 500-signature petition calling for the Council to declare the County a Frack Free Zone on the 13th October 2014.

Scotland's First Victory…
In the meantime, Reach Energy had submitted an application to drill an exploratory borehole for Coal Bed Methane in North Lanarkshire in November 2011. A leaflet and petition campaign was the response made by the local community which managed to tally up an impressive 209 formal objections and an even larger signature count on the petition. Such was the scale of the local feeling that the landowner requested that Dart Energy withdrew its application, which was formally announced on 27th February 2012.

…And One Up for Wales!
With the anti-fracking movement slowly gaining momentum across the country, the Coal Bed Methane and Underground Coal Gasification campaigns were beginning to emerge. Although the UK's CBM technology is at a more advanced stage than shale gas drilling (around 60 permissions had been granted for CBM exploration around the UK by May 2012),

the main resistance was developing around Lanarkshire in Scotland. A number of CBM wells had been permitted for Wales, but the biggest threat to the nation came from UCG – a method that has so far never been tested in the UK at all. With Clean Coal Ltd holding a license for UCG exploration under Swansea Bay, a small number of local campaigners protested outside an oil and gas industry conference in London on 3rd May 2012 in opposition to what they described as "the nuclear option" of unconventional gas extraction.

The Fracking Threat Returns

In Lancashire, Cuadrilla had been submitting a number of applications throughout 2012 for their shale gas operations. The first came on 20th December 2011 and asked permission to extend the time allowed to drill, frack and test the Singleton well until 20th May 2014. In June, the company announced its intention to drill its fifth borehole (previously permitted in 2010) the following month, though with the moratorium on hydraulic fracturing still in place, they did not intend to test-frack this well. They had also been conducting extensive seismic monitoring of the Preese Hall site to identify the natural fault lines in the underground geology, finding a 2,000 foot (609m) long fissure in close proximity to the well bore.

Nonetheless, the fear was that the ban could be lifted at any moment since the British Geological Survey report into the 2011 earthquakes recommended that fracking be opened up across the country, and a Government announcement was expected within a matter of weeks. In anticipation of this, a group of 20 activists blockaded a compound in Chesterfield owned by drilling contractor P R Marriot, where the dormant Cuadrilla rig was being stored on the morning of 18th June 2012. Locking themselves onto the gates, they aimed to stop the work of the contractors in undertaking maintenance work on the rig – nicknamed the 'Lord Browne' by campaigners after the Cuadrilla Chairman and former Cabinet Office Lead

Non-Executive, Lord John Browne – and prevent it from being mobilised to Lancashire.

A follow up event occurred on 25th July when a large gathering of local residents and campaigners came together at the windmill at Lytham St. Annes near Blackpool – touted as the 'Gathering on the Green' – to discuss thoughts and ideas on the threat posed by the five Cuadrilla well sites, and to give a unanimous 'No' to the process of hydraulic fracturing and any further exploration of shale gas on the Fylde Peninsula. Bringing together three local campaign groups:- Resident's Action on Fylde Fracking, Ribble Estuary Against Fracking and Frack Free Fylde, the event attracted significant local media coverage, helping to get their message out to the wider community.

Toxic September

In the same month that Cuadrilla applied for an extension to its operations at Banks, the Environment Agency gave the company permission to transport its contaminated produced water from its 2011 Preese Hall fracking site. 18,000 gallons of the fluids had been collected and stored in surface containers after the well was fracked in April and May 2011. In October 2011 8,000 gallons were transported the 46 miles (74km) by road to a waste water treatment works at Daveyhulme in Manchester, but the Environment Agency declared the water to be too radioactive to move and that to do so breached UK regulations. However, despite the same regulations still being in force, the Environment Agency gave special permission for the remaining 10,000 gallons of produced water to be moved by tankers to Daveyhulme in September 2012 where, after treatment and much to the distress of campaigners and environmentalists, it was expelled into the Manchester Ship Canal.

Meanwhile in Scotland, Dart Energy (who in November 2012

would make public its intention to drill an exploratory gas well just 1.2 miles [2km] from the site of the infamous Auschwitz Concentration Camp) made two applications for what could become the UK's first unconventional gas development on 6th September (Stirling Council) and 7th September (Falkirk Council), to the development of Coal Bed Methane production which would necessitate the development of 14 well pads, pipelines and other associated infrastructure. In total, 22 new wells would be drilled from the 14 sites, with around 12.5 miles (20km) of pipelines and an outfall pipe into the Firth of Forth for the disposal of treated produced fluids. Dart Energy had by this time already been operating a small scale CBM well in Cheshire, but these joint applications represented the first major unconventional gas development in the UK. At the time of writing, both applications are in the process of being appealed after Dart Energy applied to the Scottish Government on 5th June 2013 on grounds of non-determination.

All Quiet on the Fracking Front
With the situation regarding fracking still uncertain, Cuadrilla began to apply for more permissions at its Lancashire sites, temporarily removing the process of hydraulic fracturing from their plans. On 19th September, as we have already seen, Cuadrilla applied for an extension to its testing and restoration plans for the Banks well site until 28th March 2014 (application is yet to be determined). Then on the 30th November they applied to extend the time allowed to restore their St. Anna's Road site until 31st July 2014, specifically stating that there would be no hydraulic fracturing at all. For the time being at least, it would appear that Lancashire was to have a break from the threat of the frackers drill.

Arise, Sir Frackman
With the brief lull in the immediate threat to his home, St.

Anne's resident, Gayzer Frackman, took the opportunity to raise awareness of the dangers of fracking directly to the Prime Minister, David Cameron, in a rather novel way. The Childrens Entertainer embarked upon a walk from Blackpool Tower to No.10 Downing Street on 5th October 2012. His route was to take him to many of the areas under threat of CBM and shale gas exploration across England and Wales, carrying with him a petition which was added to as he visited each city, town and village he passed through. He reached Warrington in the 6th October and then Wrexham the following day. On the 8th he took a route to Shrewsbury, then to Leominster (9th), Ross-on-Wye (10th) and reached Cardiff on the 11th. Staying in Wales, he made the walk to the Llandow drilling site, before returning to Cardiff in the evening of the 12th October. Heading back in England, he reached Chepstow on the 13th, Gloucester (14th), Swindon (15th), Reading (16th) and the Surrey city of Guildford on the 17th. Taking a detour in West Sussex, he made a visit to Balcombe on the 18th in the first sign of solidarity between fracking campaigners in the north and south of England, before reaching the doorstep of the Prime Minister in Westminster the following day.

This epic journey, all to raise awareness of the fracking threat to the UK, made the 51-year-old Gayzer, dressed in his iconic colourful outfit, a famous figure in the anti-fracking community.

The Big Rig Revolt

The 1st December 2012 will be a day that will forever live in infamy amongst the UK anti-fracking community. This day marked the first ever national day of anti-fracking action, with hundreds of people taking part all over England, Wales and Scotland. It also marked a wider, global day of action on climate change as delegates from the United Nations were halfway through another attempted round of climate talks.

Kicking off the day, the Extreme Energy Tour began collecting letters of support at 10am for a Parliamentary Motion calling for a complete moratorium on fracking with a 'Clean Energy not Extreme Energy' stall at Huddersfield's Market Place before conducting a walking tour of the High Street visiting businesses and shops owned by companies known to be financing or otherwise facilitating fracking and other unconventional oil/gas extraction around the world. At each stop, a 'tour guide' explained to the crowds how each business was involved in the fossil fuels industry, before a protest letter was delivered to the various shop managers.

Next came a protest in Kingsmead Square, Bath - by Frack Free Somerset against UK Methane's application for Coal Bed Methane exploration near Keynsham at 11am. This took the form of a mock fracking crew in hard hats and high visibility clothing erecting 'fracking rigs' at various locations around the historic city and engaging with passers by. Simultaneously in Swansea, more than 100 members and supporters of Safe Energy Wales conducted a protest against Europe's first potential Underground Coal Gasification site at Swansea Bay by cordoning off an area in front of the castle where they also erected a drill rig and displayed an array of signs showing the plans for the region. People were polled for their opinion on whether they would support the 'frackers' in continuing to 'drill' under the castle and expanding out to other areas around the Bay.

In Lancashire, Ribble Estuary Against Fracking and Residents Action on Fylde Fracking embarked upon a protest march at midday from the St. Annes train station to the St. Anna's Road drilling site, intended to coincide with a protest occurring in London.

The final of the regional events was organised by Sussex Extreme Energy Resistance at 1pm in New Road, Brighton where there was another large protest with many in fancy dress related to the subject of oil and gas. A mock fracking rig was also erected and people were engaged on the subject as they passed by. Brighton was the hub of the emerging south-east anti-fracking movement, with the City Council having also passed a motion declaring Brighton and Hove a 'Frack-Free Zone'.

The main event, however, was a 1,000-strong National Climate March beginning in Grosvenor Square, London at midday. With the crowd assembling, a 656 feet (200m) long mock-up of the Keystone Pipeline threatening areas of the northern USA and Canada was 'constructed' between the Canadian High Commission and the Embassy of the United States of America. This was followed by a 23.5 foot (7.2m) tall drill rig assembled in front of the Houses of Parliament. Additionally, at 8:30am, a number of Frack Off activists erected a 20 foot (6m) tall drill rig outside the home of Lord John Browne. A contingent of the National Climate March, headed by Vanessa Vine from Sussex and supported by representatives from Falkirk, Belfast, the Fylde and Ribble Estuary, and the Vale of Glamorgan handed in a protest letter to David Cameron at 2pm.

Lancashire – New Year, New Fracks?
Returning to Lancashire, Lancashire County Council summarised the situation in late 2012 as:

> "Planning permission has previously been granted for a number of temporary exploratory drilling operations for shale gas on sites including sites at Annas Road and Becconsall [Banks]. The planning permissions to these sites provided for 'fracking' and were time-limited in view of their temporary nature. The Annas Road and Becconsall sites have been developed; an initial borehole at Annas Road had to be abandoned at

shallow depth due to technical problems and it is proposed to drill a second borehole; the borehole at Becconsall has been drilled although no 'fracking' has taken place. As a result of the suspension of 'fracking' activities by the Department of Energy and Climate Change, the applicant has applied for extensions of time to complete the exploratory operations. An application has also been received to drill a horizontal borehole from the vertical at Annas Road. The applications at Annas Road do not provide for 'fracking'. The principle of shale gas exploration and the applications have raised a number of issues."

Still intending to test frack at Banks, the Annas Road site had at least been spared for now. However, on 11th January 2013, less than two months after ruling out hydraulic fracturing, Cuadrilla submitted a **Scoping Opinion** to Lancashire County Council to accompany an application for the "testing and hydraulic fracturing of exploratory lateral borehole" at the Annas Road site. Confusingly, the aforementioned application, validated just five days later, expressly ruled out any fracturing at the site, instead wishing to take a core sample of the shale rock along a lateral borehole.

More Eyes on Sussex

On 20th July 2012, Celtique Energie became the second oil and gas exploration company to set its sights on the Weald Basin of Sussex. With the well site at Balcombe not having yet started, the subject of hydraulic fracturing was still largely unknown to large parts of West Sussex, although Celtique had most likely seen the activity in Lancashire developing into a more widespread movement, and so it was made clear that there would not be any hydraulic fracturing proposed in this application. Instead, the target reservoir was the Sherwood Sandstone layer which would be naturally porous and allow any oil or gas to flow freely. Celtique stated that they wanted to explore for hydrocarbons – expected to be gas – in the Willow Prospect, an identified reservoir with potential for economic production. The selected site for the drilling was Woodbarn Farm, near the tiny hamlet of Broadford Bridge and located approximately halfway between the large villages of Billingshurst to the north and Pulborough to the south. With little concern raised by the local population (3 objections were registered and 2 further letters of concern) and no objections raised by West Chiltington, Billingshurst or Pulborough Parish Council, and assurances of conventional reservoirs and conventional drilling methods, West Sussex County Council granted permission on the 11th February 2013.

However, campaign group Frack Off revealed in May 2013 that the Sherwood Sandstone targeted by Celtique lies approximately 1,000 feet (304m) below the Liassic Shale layer, meaning that Celtique will have to drill through it to reach the 'conventional' target reservoir. Furthermore, a deleted section of Celtique's own website reveals that the Willow Prospect is congruent in part to the Weald Basin shale prospect and that the company is keen to explore the shale potential of the Weald Basin. Also, Celtique's investment partner and co-owner of the Petroleum Development and Exploration License, Magellan Petroleum, announced in a letter dated 28th May 2013 to its shareholders that the Weald Basin "is a very promising unconventional play" and that it is "one of only three publically traded companies to offer significant exposure to this emerging UK shale play". More importantly, the letter also states that "together with our partner Celtique Energie, [we] plan to drill one or two evaluation wells at the end of 2013, through which we will gain a better understanding of the shale potential of our acreage". Magellan Petroleum describes itself as "an independent oil and gas exploration and production company focused on…the exploration of unconventional hydrocarbon resources in the Weald Basin".

The Battle of Balcombe

Just as Celtique announced its intentions for future applications to West Sussex County Council later in the year for sites at Wisborough Green (4.5 miles [7.3km] north-west of Broadford Bridge) and Fernhurst, in the South Downs National Park (13.5 miles [22km] north-west of Broadford Bridge), Lancashire fracking company Cuadrilla prepared to move onto its Balcombe well pad – marking the beginning of what became an historic and groundbreaking summer. The UK Anti-Fracking movement was soon to receive international media attention.

Fig.12: The drilling begins at Balcombe, July 2013

For 68 incredible days, the small West Sussex village of Balcombe, located deep within the Conservative heartland, became the front line in the fight against fracking and fossil fuels, and the focal point for opposition to the Conservative-led Government's energy policies. The Great Gas Gala officially began on the 25th July 2013, when Cuadrilla attempted to begin delivering equipment to the site, located at its closest point just over half a mile (1.1km) from the Ardingly Reservoir that supplies over 70,000 homes with clean drinking water. A small number of local residents, knowing that the deliveries were due to commence, had been waiting outside the entrance gate in preparation for when the first lorry arrived at around 8:30am. As more lorries arrived, more people took part in the blockade so that by midday over 250 people were present. Knowing that the only hope was stopping the equipment from entering the site, several tents and gazebos were erected on the grass verges so that some could stay overnight – marking the momentous turning point when an otherwise ordinary protest transformed into the UK's first community blockade, a situation that would be replicated later in the year and throughout 2014 at other drilling sites across the UK. On day 2 (26th July) the police presence increased and the first attempt to break the blockade was made at 12:30pm. By 1pm, 10 **Protectors** had been arrested under Section 241 of the Trade Union and Labour Relations Act 1992 for allegedly besetting and intimidating the site workers. A meeting was held by the protectors and it was decided that the blockade would continue despite the arrests. Day 2 also saw the first police escorts to the arriving lorries – a scene that would be replicated almost every day of the blockade. A cycle of more police, more lorries, more protectors and more arrests continued. The police were heavily criticised during and after the event for the tactics that were employed and for the high number of arrests at what was a peaceful protest – there were no reported incidents of violence or threats during the whole of the 68 days.

In total, 126 people were arrested, with 114 having charges made against them. However, in the end, only 26 protectors were found guilty of having committed any offence. Many were found to have been unlawfully arrested or not enough evidence was provided to prove an offence was actually committed. The bill for the policing of the protest amounted to around £4 million – equating to £31,746 per person arrested, or a staggering £153,846 per person actually found guilty of an offence. The most high profile arrestee was Caroline Lucas MP, charged with obstructing the highway but found not guilty in court.

The Balcombe Blockade also marked several other important factors in the emerging anti-fracking movement. From the very start of the blockade, the UK Chapter of the Knitting Nanas Against Gas have been an ever present and much loved sight at Balcombe and all other anti-fracking events.

Originating in Australia, the Knitting Nanas protest against oil and gas extraction in a highly unique way – by peacefully and quietly knitting hats, scarves and other items outside protest sites, offices of the drilling companies, court appearances and at the offices of pro-shale local politicians. Furthermore, national media attention was cemented when the action group No Dash For Gas decided to make a last minute relocation from a planned encampment at EDF's West Burton Power Plant to the Balcombe Blockade in August 2013. Lasting from 16th to 21st August, the relocation was in response to a call out by Balcombe residents for more help to man the blockade. For the first two days the 'Reclaim the Power' camp provided a focal point for new arrivals to the site, with several areas dedicated to food, awareness, children and direct action training. With the support of the vast majority of the Balcombe residents, the proposal for large scale and co-ordinated direct action against Cuadrilla was welcomed. After attempts to stop the drilling through the planning system, via petitions, letters to MPs and the Prime Minister had all failed, this method of peaceful 'civil disobedience' was seen as the only option left. On the Sunday, 18th August, a mass protest march outside the site was organised, attracting over 2,000 people from across the UK – by far the largest show of anti-fracking solidarity seen so far.

Fig.13: The Knitting Nanas at the Balcombe Blockade, Summer 2013

The programme of direct actions began on 19th August with members of No Dash For Gas and Frack Off managed to shut down Cuadrilla's Lichfield offices and their public relations contractors, Bell Pottinger, in London. At the Bell Pottinger offices, six activists glued themselves to the doors and to each other, whilst others erected banners. A sound system was also used to play an alleged recording of a Bell Pottinger employee admitting that their pro-fracking PR campaign "sounds like utter f*****g b******t". Shortly afterwards, a number of activists climbed onto the roof of Horsham Constituency MP, Francis Maude's office in Horsham, West Sussex, where more banners and a wind turbine blade were displayed. Maude, whose constituency includes Balcombe, is accused of being overtly pro-fracking and was responsible for appointing the Cuadrilla Chairman, Lord John Browne, to the post of Lead Non-Executive in the Cabinet Office. At Balcombe itself, the welfare campaigners Disabled People Against the Cuts organised a complete blockade of the site. Five activists locked onto the site entrance gate and were surrounded by a second line of blockades and a tertiary blockade of the surrounding road. A double-decker bus full of children toured the area with a large banner urging the Government "Don't frack with our future", whilst hundreds of people across the UK flooded Cuadrilla's phone lines with calls. The home of Lord Howell – father-in-law of Chancellor George Osborne – was also picketed in protest against his comments that the south-east should be spared from fracking, but that it should go ahead in "the desolate north". Lord Howell's comments, far from encouraging the anti-fracking movement, gave birth to The Force From the North – a union of anti-fracking activists from the north of England that became long-term 'residents' at the Balcombe protest site. Whilst entirely peaceful, the day of action attracted significant police attention, and saw the largest number of arrests for a

Fig.14: The Balcombe Blockade, Summer 2013

single day, though most charges were later dropped. In one of the most pitiful legal actions against the activists, one person who was involved in the blockade of the Bell Pottinger offices on the 19th August was found guilty in court and was fined for £4 worth of criminal damage and banned from carrying glue in public. Although the Reclaim the Power camp officially ended on the 21st, many of the participants remained at Balcombe, seeing the protest through to its end on 30th September 2013.

Balcombe was highly significant in the anti-fracking story of the UK. Whereas before the media attention on the movement was largely confined to local press during public events, the world's mainstream media swooped on the tiny village – helping to spread the message of the dangers posed by hydraulic fracturing and other unconventional fossil fuel extraction. A number of famous names also lent their support to the protectors, with many joining in the protest at various times. Bianca Jagger, Natalie Bennet, Keith Taylor MEP, Caroline Lucas MP and Dame Vivienne Westwood were amongst those taking the time to visit the site. Whilst some media outlets attempted to show the anti-fracking movement in a negative light, the amount of attention given to the protest had the opposite effect. For the first time, fracking was a word known to almost every household in the country. People were beginning to talk about fracking as a national issue. In the immediate legacy of the Balcombe Blockade, at least 26 public meetings were advertised across England in the month between the 25th September and 25th October, including several council and parliamentary debates. Many of the events had to arrange last minute venue changes since the interest shown in attending was so great, whilst others were forced to be ticket-only events. Furthermore, a large-scale survey by the University of Nottingham revealed that post-Balcombe, the amount of public support for shale gas exploration was dropping considerably. Until as recently as September 2013 most polls showed that the pro-fracking lobby, even at its lowest performance, was level, if not slightly higher, than the anti-lobby. For the first time, the number of people opposed to hydraulic fracturing outnumbered those in support of the technology. For many, the Balcombe Summer of 2013 marked the true birth of the UK anti-fracking movement as a national issue.

Towards the end of the protest, West Sussex County Council made the decision in September to evict the protest camp and take possession of the roadside around the drilling site, despite having received a number of complaints from residents about alleged breaches of planning permission. On the 16th September, the Council took the camp to the High Court, but the judge ruled that they had failed to take into consideration the human rights of the protectors regarding the lawful right to peaceful protest and granted the camp permission to remain on site until 28th September, when Cuadrilla was due to complete its operations. On the 30th, the camp had largely dismantled, but announced that it would clear the site by the 8th October. In advance of preparing another legal challenge, West Sussex County Council, in association with Sussex Police issued an order for the creation of a 'protest pen' opposite the site entrance, which would permit limited protest but ban the use of tents and prevent any form of protest outside of the designated area. This limitation was removed in a legal challenge mounted by the camp, since it infringed the right to assembly and peaceful protest. Another High Court appearance was announced by the Council on 8th October to evict the camp, and on the 11th November, the High Court ruled in favour of the Council. By this time, only a very small number of people had remained camping on the roadside, citing that they were there to ensure that Cuadrilla didn't attempt to re-enter the site. However, the police and bailiffs forced their vacation. In response to the Council's actions and to claims that their views were not being listened to, around 25 members of the former Balcombe Blockade set up a small protest encampment on the lawn of the Council's headquarters at County Hall in the city of Chichester on 17th November, in an act to force discussions with the Council. The camp lasted for just the weekend and on the Monday, the leader of the Council, Louise Goldsmith, announced that she would be willing to host talks with anti-fracking campaigners from groups across West Sussex. For a day-by-day report of activities at the Balcombe Blockade, see the Frack Off website.

Fig.15: Anti-Fracking campaigners camp on the lawn at West Sussex County Council's County Hall in Chichester, 16th November 2013

Before the protests, Cuadrilla had also applied for two additional planning permissions for Balcombe. The first, received by West Sussex County Council on the 1st July 2013 was for additional time to conduct flow testing and the second, received on 17th July was to change the type of flare to be used. Hundreds of objections were registered for these two applications – contrary to the earlier application in 2010 – and in a twist of fortune, Cuadrilla decided to withdraw both applications on the 2nd September 2013. This, however, was not the end of the activity at the tiny West Sussex village. Furthermore, Celtique Energie, who received permission to drill at Broadford Bridge, 14.4 miles (23.3km) south-west of Balcombe, in February 2013 had also submitted another application to West Sussex County Council for an exploratory well at Wisborough Green on the 6th September 2013 and a Scoping Opinion to the South Downs National Park Authority for a site at Fernhurst on the 7th June, with an application following on 11th December 2013.

England's Green and Pleasant...Water?

In a bizarre event, villagers at Balcombe reported that sections of a stream running past the Cuadrilla drilling site had begun to turn bright green. The environment Agency conducted on site tests and found that the unknown substance was harmless, but the source was unknown. Agricultural products were ruled out after chemical analysis showed there was no ammonia present and that oxygen levels were fine. More detailed laboratory testing also proved inconclusive, but the Environment Agency suspected that it was a green dye that had entered the watercourse. Supporters of Cuadrilla immediately accused the activists of intentionally releasing the substance into the stream, though this was denied by all parties. The source and cause of the temporary discolouration remains unknown at the time of writing, though one theory is that it could have been a harmless dye known to be used by the drilling industry to check for leaks from wells. Was it a hoax, or is this evidence that the Balcombe well had failed?

Double Trouble for Barton Moss

On 17th June 2010, Salford City Council granted Igas permission to drill two exploratory boreholes for Coal Bed Methane at Barton Moss, on the outskirts of Manchester. The application attracted 138 comments – a sizeable number for a pre-Balcombe application! Whilst the primary target was the coal layer, deeper drilling was planned to sample the shale and other tight formations.

Following a delay in attaining permits from the Environment Agency, and with Balcombe still fresh in the minds of oil and gas operators, Igas began to prepare its site ready for the rig to arrive in mid-November 2013 by erecting extra layers of security fencing. Public meetings to discuss what action should be taken were held in the run up to the anticipated drilling date, whilst a protest organised by Say No to Fracking on Barton Moss took place outside an 'Igas Open Day' at Salford City Stadium on 15th September, and so a repeat of the Balcombe Blockade was expected. The Northern Gas Gala – as it was being advertised – was waiting in preparation.

Fig.16: The IGas drilling rig at Barton Moss, December 2013

In total, the Barton Moss Blockade lasted for 137 days between 27th November 2013 and 12th April 2014, more than twice as long as the Balcombe Blockade, and the longest continuous anti-fracking protest in the UK to date. From the outset, the police presence was enormous and at numerous times was accused of being violent and overtly aggressive against the protectors – several people were injured during arrests, requiring hospital treatment. At the conclusion of the blockade, a legal review into the tactics employed by Greater Manchester Police, and in particular their Tactical Aid Unit, was announced and the results of which are awaited with much anticipation. Numerous official complaints have also been logged with the Independent Police Complaints Commission. Knowing also that the main point of the blockade was to prevent access to the site, Greater Manchester Police utilised a system of convoys to force a gap in the protest and allow as many lorries as possible to enter the site in one go. Nonetheless, and learning from the experience at Balcombe, the protectors continued to walk as slowly as possible in front of the lorries in order to delay them, thereby costing the drilling operators heavily. It was during the Barton Moss Blockade that the protectors began to refer to themselves as the Investor Removal Team. Whilst it has become evident that police forces up and down the country will not permit complete blockades of drilling sites, the tactic is to cause as much delay and disruption as possible, so that unconventional oil and gas extraction in the UK fails to be an attractive option for financial investors and becomes uneconomical for the drillers to continue. Every extra minute that it takes a lorry to make a vital delivery, makes the costs of the operation rise significantly.

Figs.17 and 18: The Barton Moss Community Protection Camp and delivery traffic competed for space on the narrow access road

On the night of 4th to 5th December 2013, high winds battered the camp, causing a lot of damage to the tents and other temporary structures, though thankfully with no injury. With help and assistance – financial and physical – from people all across the UK, the camp was rebuilt on the 10th. On the 13th December, a convoy of 12 lorries, under heavy police escort, were successfully delayed for two hours. The police received further widespread condemnation after a disabled man suffered a broken knee after being knocked to the ground. Three days later a team of 50 activists managed to assemble an impressive 54 foot (16.5m) long, 1.5 tonne (1,500kg) wind turbine blade across the site entrance in the early hours, completing the assembly and vacating the site before the police could arrive to stop the work. Requiring a large team and heavy equipment to dismantle and remove the early Christmas gift –

the activists kindly tied a festive red bow around the object – the site was completely blocked for many hours, preventing any of the scheduled deliveries from taking place. This was followed on the 18th December by a large orange bus that had "mysteriously broken down, blocking the entrance to the fracking site". Five protectors then proceeded to lock onto the bus, preventing its removal by the police. This echoed an event at Balcombe earlier in the year where a large fire engine was parked across the site entrance and locked onto by a number of activists, symbolising the amount of water that fracking operations require. At the bus protest, one protector locked onto the steering wheel and accelerator and another attached themselves to the underside of the vehicle. A third had locked onto the skylight on the roof whilst a fourth attached themselves to the back door by their neck. The fifth person locked their leg onto the front door. This arrangement meant that the police could not gain access into the bus or in the event that they did, that they could not drive it nor tow it away. It took over six hours before the bus could eventually be removed which caused severe disruption to the planned arrival of the second drilling rig onto the site that same day.

Following the number of complaints to the police about alleged aggression during their policing efforts, a noticeable restraint was felt by day 24, and the protectors reported that there was less tension and fewer arrests occurring. Despite Christmas Day being a traditional time of spending time at home with the family, the camp held strong throughout the holiday season, and many more people from around the country travelled to the site with deliveries of food. Local businesses in Manchester supplied dozens of mince pies to the protectors whilst a member of Residents Against Fylde Fracking cooked an entire Christmas Lunch. On 30th December, a familiar face from Balcombe managed to get past the police and locked himself onto the gate. Jamie 'Tripod' Spiers came to fame amongst the anti-fracking community earlier in the year by attaching himself to a tall tripod-like contraption that blocked the road outside the Cuadrilla site for several hours.

In a shocking event on the 6th January 2014, Greater Manchester Police conducted a tent-by-tent raid of the community blockade after they alleged that in the early hours of the morning, a person had fired a flare at a police helicopter hovering over the site. This was hotly denied by all present and the search revealed no evidence of any flares or related paraphernalia, though much of the contents of the tents had been deposited into the mud and water, causing much distress and damage to possessions. No one has ever been charged or convicted over the alleged offence and Greater Manchester Police have refused to meet with protectors to discuss the event. Six days later hundreds of activists from across the UK travelled to Barton Moss in a united show of defiance against the shale industry. The following day a united day of protest resulted in a mass of people blocking the end of the Barton Moss road, preventing an entire convoy of tankers and lorries from getting near the site, whilst a number of protectors scaled onto the queuing vehicles and proceeded to 'occupy' them for as long as possible. In another symbol of the environmental damage caused as a result of fossil fuel extraction, police escorted tree surgeons on the 22nd January to remove all the trees lining the lane leading to the Igas site. No explanation was offered and two suspected reasons have been proposed. The first is that an observation post had been constructed in one of the trees to keep a watch over the activities going on a the drilling site and that the felling of all the trees was an attempt to remove this post and prevent any more being constructed. The second speculated theory was that it was in preparation for the widening of the road to allow the lorries easier access to the site. In fact, at a later date a court ruled that the 'road' so many people had been arrested for blocking was in fact a footpath and not a road at all. This means that all the arrests had been unlawful and that the driving of motorized vehicles along the footpath was also illegal. However, none of the charges were dropped in the end. Nonetheless, on the 13th February, the day after the court ruling, an all-day blockade occurred for the first time, preventing any vehicles entering the site, since the police could no longer arrest the protectors for obstructing the highway. As with Balcombe, the remainder of the encampment was largely a cycle of protests, blockades, arrests and delivery convoys.

Another notable event was Solidarity Sunday on the 26th January. In a repeat of the previous Solidarity Sunday in August 2013 at Balcombe, buses were arranged to transport people from 15 different locations around the UK, including Blackpool, Bristol, Balcombe, London, Swansea, Falkirk, Leeds and Sheffield. Then on the 11th February, a convoy of lorries making a delivery to the site were found to be transporting radioactive material, causing much concern about the activity going on behind the security fences. For a day-by-day account of activities throughout the 137 day community blockade, visit the Northern Gas Gala website. In early 2014,

Figs.19 and 20: The felled tree has become a symbol to the anti-fracking movement

Bez from the Happy Mondays announced at the Barton Moss Blockade that he was to stand in the 2015 General Election for the constituency. He also arranged for financial expert Max Keiser of Russia Today's The Keiser Report to address a protest arranged in the centre of Manchester, bringing more famous names into the anti-fracking movement.

Aside from the duration of the encampment, the largest notable difference from Balcombe was the distinct lack of national media attention. Although occasional pieces entered the mainstream press, for the majority of the time there was little coverage of what was one of the largest and longest mass protests in recent history. It is a widespread belief that this sudden change in attitude of the mainstream media had much deeper roots than a simple lack of interest. After all, there had been some potentially massive news stories emerging out of Barton Moss. With the Government and industry recognising just how large an impact the Balcombe Blockade had been upon public opinion – in particular how it inspired thousands of people across the country to take up non-violent direct action as a legitimate form of protest against large, wealthy and incredibly powerful entities – it is held by many that a degree of political infiltration of the larger media outlets is attempting to reduce coverage of the Frack Free Movement. Whether this is true or not will probably never be known, but the rapid u-turn made by the popular press after the Balcombe Blockade cannot help but look suspicious, particularly when it has been negative – or some would argue, manipulative – articles that made up the bulk of the mainstream media coverage on the few occasions that a news story broke out.

Winter 2013 – Spring 2014
On the 4th October 2013, Cuadrilla announced via the Blackpool Gazette that it's beleaguered drill site at Anna's Road in Lancashire – previously troubled by technical issues in 2012 and a second borehole was proposed – had once again

suffered from problems. The company was forced to pull out of its operations after "technical constraints related to wintering birds" meant that drilling was only permitted for 6 months of the year.

A few months later, on the weekend of the 18th and 19th January 2014, Cluff Natural Resources attended a public meeting arranged by local residents to discuss its future plans for Underground Coal Gasification under the Loughor Estuary near Llanelli in Wales. The company received a license for the UCG project in 2013 and propose a land-based operation to drill into the coal layers under the estuary. No application has yet been submitted as of the time of writing, but Cluff Natural Resources are undertaking a series of feasibility studies. Around 100 people attended the meeting and left the company in no doubt that its plans were not welcome.

Meanwhile, the Coventry Telegraph announced on the 23rd January 2014 that the same company had been engaged in "highly secretive" confidential discussions with senior Council members at Warwickshire County Council about potential plans to explore for Underground Coal Gasification opportunities in the county. The company does not currently hold a license for the region, but is in the process of bidding for an area of land about the same size as the city of Coventry in the Warwickshire countryside. If permitted, this would be the first onshore UCG license for the UK.

Falkirk in the Spotlight

Returning to the Dart Energy proposals for 22 Coal Bed Methane wells at Airth, near Falkirk in Scotland, we have already seen that on the 5th June 2013, two appeals were lodged with the Scottish Directorate for Planning and Environmental Appeals (DPEA). Although both appeals relate to the same project, since the proposed gasfield development crosses over into the Stirling Council area, Dart had to submit two separate, but related, planning applications – one to Falkirk and one to Stirling. Over 2,500 objections had been registered at the time of the appeal, and after being pressed for a response by the DPEA, both councils also made objections to the proposals. The public inquiry began on the 18th March 2014 and was challenged by an alliance of local community councils and other residents known as the Concerned Communities of Falkirk. In total, seven Community Councils and a further 18 West Fife coastal villages came together under the alliance, with a combined representation of over 90,000 residents! The Inquiry lasted until the 2nd April, with the closing statements by both parties held on the 15th and 16th of April. The results of the decision were due to be announced in June, but changes to the Scottish Government's planning policies led to further written representations being made on the 23rd of that month. Then, on the 11th August, the DPEA granted Concerned Communities of Falkirk the chance to make further submissions based upon Dart Energy's closing statement in April. This meant that Dart and the Scottish Environment Protection Agency were given until 26th August 2014 to respond. The result of the DPEA's decision has yet to be announced at the time of writing, and is eagerly anticipated. Dart Energy's plans represent the first gasfield development in the UK, and if permitted would also become the largest production site for unconventional energy.

Secret Shale

Early to mid-2014 also saw a wave of other sites across England come to the fore as being not all that they appeared to be. We have already seen that Celtique Energie received permission to drill a conventional oil well near the hamlet of Broadford Bridge, between the two West Sussex villages of Pulborough and Billingshurst, but that upon closer examination revealed that whilst the targeted reservoir was indeed 'conventional' the borehole would be drilled through the shale layer of the Weald Basin. Additionally, a press release by Celtique's American investment partner, Magellan Petroleum, dated the 3rd June 2014 announced that the well "will be drilled vertically and completed [later in 2014] without the use of hydraulic fracturing and ultimately target conventional Triassic gas plays. However, during drilling, Magellan will have the opportunity to core and log various shale and tight formations in the Cretaceous and Jurassic sections of the Weald Basin". In the same press release, Magellan make the same statement for another 'conventional' well at Horse Hill, near the village of Hookwood in Surrey (commenced in early September 2014), for which they received permission from Surrey County Council on the 9th November 2011. Furthermore, in an Investor Presentation by Magellan Petroleum in early 2014, one slide entitled "UK Unconventional" a specific reference is made to the Broadford Bridge exploratory well site.

Fig.21: The drilling rig at the aptly-named Horse Hill site in Surrey, 4th October 2014

Remaining in West Sussex, and Celtique also submitted two further applications for a site between the villages of Kirdford and Wisborough Green (application received 6th September 2013) and a second site near the village of Fernhurst (received 11th December 2013) in the South Downs National Park. The Kirdford/Wisborough Green application proposed to drill to a depth of 8,750 feet (2,667m) with a combined vertical and lateral well. The lateral well, branching off the vertical at between 4,000 and 5,000 feet (1,219m to 1,524m) would target the Kimmeridge Limestone layer, which was described by the company on their public exhibition boards as being a tight formation linked to the oil/gas-producing shale layers (though elsewhere Celtique claimed it was a conventional formation), whilst the vertical section would continue down and into the Liassic Shales. A massive local campaign ensued almost as soon as the proposals were first announced, and a series of public meetings were held in Wisborough Green.

At one stage, a mass protest march was held in October 2013 by two hundred residents from both villages that converged at the proposed drilling site. Additionally, taking inspiration from the success of a similar move by landowners at Fernhurst, nine landowners neighbouring the proposed site announced on the 21st July 2014 that they had formed a legal blockade, expressly denying permission to drill under their land and limiting the ability of Celtique to undertake their operations. As it happened, West Sussex County Council received 2,471 objections to the application, to only 18 of support and on the 23rd July voted unanimously to refuse permission – becoming the first Council in England to oppose an oil and gas application. Amongst those opposing the plans were the well known actors James Bolam MBE and his wife Sue Jameson, and the Conservative Member of Parliament for the Arundel and South Downs Constituency, Nick Herbert. As with the nearby Broadford Bridge and the Horse Hill applications, Celtique insisted that it would have been a conventional well, though both targets were tight formations. They repeated that there would not have been any hydraulic fracturing, although future use of the technology in the vicinity was not ruled out depending on flow testing. An appeal against the Council's decision was made shortly before the time of writing and a date for a public inquiry set for October 2015.

Due to the sensitivities of the Fernhurst site – lying as it does in the South Downs National Park – Celtique undertook a two-stage application process. A Scoping Opinion was submitted to the National Park Authority on the 7th June 2013 (receiving 119 public comments), followed by the full application on the 11th December 2013. A record breaking number of formal objections had been registered for this application, totalling 5,517 by the time the decision was made. As with Wisborough Green, Celtique requested several postponements

Figs.22 and 23: The proposed site at Wisborough Green is just behind the trees. The access road would follow the dirt track visible. All vehicles serving the site would be forced to cross the old and very narrow Boxal Bridge

to the decision date in order to make a number of amendments, and so the South Downs National Park Authority announced in late August that the date had finally been set for the 11th September 2014. This application presented perhaps the most strategically important decision for the future of oil and gas, particularly unconventional exploration and exploitation, in the UK as a whole, since it was the first and so far only application proposed for a National Park – the highest level of landscape designation and protection available. If permitted, the precedent would essentially open up almost the entire UK as a free-for-all to further applications. The tension amongst the local community was raised when, on the 26th August 2014, the

West Sussex County Council Highways Department withdrew their objection to the application following additional information supplied by Celtique. A week later, the South Downs National Park Authority finally released their Committee Report in advance of the meeting which recommended a refusal of the application on the grounds that "the applicant has failed to demonstrate exceptional circumstances exist for such exploration and appraisal to take place within the protected landscape, and that it is in the public interest to do so." The decision came on the 11th September and the Committee voted unanimously to refuse permission, reflecting the fact that there were no exceptional circumstances that would require the drilling for oil and gas in a National Park and agreeing that the impact of the works traffic, visual impact of the rig and the industrial development in the countryside are not compatible with the Park's purpose. At the time of writing, Celtique have announced that they are to appeal the decision by West Sussex County Council to refuse permission to drill at Wisborough Green and campaigners are awaiting the decision as to whether an appeal will be lodged for the Fernhurst site.

Among the first actions of the local community in Fernhurst was the 'Big Balloon Bash' in mid-September 2013 where a barrage balloon, displaying No Fracking banners, was tethered to ground near to the proposed site and flown at a height equal to that of the proposed drilling rig, to demonstrate how visible it would be to large parts of West Sussex and the South

Fig.24: A placard at the Fernhurst decision meeting on 11th September 2014 reminding Celtique Energie of the unprecedented level of opposition to their plans

Downs National Park, particularly when lit at night. Celtique dismissed the 'stunt' as not accurately representing a drilling rig, much to the amusement of the campaign group Frack Free Fernhurst.

Fig.25: *The Big Balloon Bash, 21st September 2013. Note the car and person at the base of the tether for scale. The proposed drill site lies immediately behind the trees*

The most momentous event that has had nationwide-impact and is believed to have incited the Government to amend its planning policies was the first ever legal blockade in the UK, where all surrounding landowners have expressly denied Celtique permission to drill under their land, or use their land to access the site (the proposed access road is on land hotly disputed by two adjacent landowners). The legal blockade was in response to a nation-wide campaign by Greenpeace, known as Wrong Move, which encourages homeowners to sign up to a collective legal block across the country. The extent of Fernhurst's blockade meant that Celtique were forced to submit an amended application in which they removed the proposed lateral section of the well, leaving them with just a single, vertical borehole. In their initial plans, Celtique proposed a similar operation to Kirdford/Wisborough Green, with a deep vertical well drilling into the liassic shale layers and a lateral well branching of along the tight Kimeridge Limestone layer.

The success of Fernhurst's legal blockade and the media attention that it attracted are believed to be one of, if not the main reason why the Government, under pressure from the industry, proposed to amend planning policy with regards to ancient trespass laws as a part of the Infrastructure Bill currently passing through the House of Commons (despite almost the entire section headed 'fracking' being left blank). Currently, landowners can claim ownership of the ground beneath their land to the centre of the Earth, but the Government proposes to remove this right for deeper than 984 feet (300m), allowing companies to drill under land and homes without the need to gain permission. Despite over 45,000 addresses across the UK having signed up to Greenpeace's legal block campaign and a YouGov survey revealing that 74% of the population oppose such changes to the law (including a large percentage of those supportive of fracking), the Government is keen to make the changes before the 2015 General Election, and both the Labour and Liberal Democrat parties have announced support for the changes also. A key argument by the Frack Free Movement is that the proposed changes are simply the Government giving in to industry pressure, not least to Cuadrilla who's Chairman, Lord John Browne was the lead non-executive in the Cabinet Office. This claim appears to be somewhat supported by comments made by Cuadrilla CEO, Francis Egan, in May 2014 with regards to the proposed law changes:

> "If you can't get [underground] access at all, if there's no amount of money people are interested in, then the resource can't be developed...I don't think companies will invest [in shale] if they think it will take years to drill each horizontal well [by taking landowners to court under current regulations]".

Meanwhile, in the north of England, Rathlin Energy had submitted applications for two sites in the East Riding of Yorkshire. The first was at Crawberry Hill, near Walkington (granted permission 18th September 2012 with 4 registered objections) and the second at West Newton near Great Hatfield (granted 17th January 2013 with 3 registered objections) – both sites around 9 miles (14.5km) from the centre of Hull. The two sites were almost identical with respects to the proposed operation and drilling depth, with a single vertical well extending down into the Lower Carboniferous Limestone layer at around 10,544 feet (3,214m). As with the three Celtique sites in West Sussex and Magellan's site at Horse Hill in Surrey, Rathlin claim that there would be no hydraulic fracturing of these exploratory wells, and that each would be entirely conventional. However, closer inspection of the application documents for the two Rathlin Energy sites reveal that the target 'conventional' reserve is in fact a tight formation that lies immediately underneath the Bowland Shale layers that Cuadrilla had been exploring (and partially fracked) in Lancashire. Additional concern regarding these two sites came to light when an inspection of the Environment Agency permit application documents (still available on the Rathlin Energy website at the time of writing) reveal the detailed drilling plans. These permit documents do not form part of the decision making process by a **Minerals Planning Authority**. At paragraphs 3.2.1 'Groundwater Activity' and 4.3.2.1 'Mini Fall-Off Test within Upper Visean/Lower Namurian', Rathlin states that this target formation, at a depth of around 10,023 feet (3,055m), "has extremely low permeability and requires mechanical intervention to enhance its permeability" and that a mini fall-off test is required to gather data about the formation. A mini fall-off test involves the perforation of the well casing and injection of fluid into the wellbore at a gradually increasing pressure to reach at or near the fracturing point of the rock. The well is then sealed off and monitored to analyse the effect of the pressure and how it is dissipated over an extended period, typically 14 days, as well as how the geology reacts to the high pressures. This test is sometimes known in the industry as a Diagnostic Fracture Injection Test, or more commonly by the wider community as a 'mini-frack'. Although Rathlin Energy, in their permit documents, make it clear that they do not intend to actually cause the rock to fracture at Crawberry Hill or West Newton, but rather "to establish if and at what pressure the formation becomes permeable...the mini fall-off test will help determine whether the formation is capable of being hydraulically fractured". Whilst hydraulic fracturing at these two 'conventional' wells (and the other sites in Sussex and Surrey) is currently ruled out, it is evident that exploration of the shale layers is a key objective that could very likely pre-empt future fracking operations at or near these locations.

2014: Five Blockades in Four Months

As the Anti-Fracking Movement grew in strength and support and spread to more and more parts of the UK since the Balcombe Blockade in Summer 2013, a wave of community resistance spread across the Midlands and North East of England in response to the unwelcome drilling activities of Dart Energy and Rathlin Energy. The first of these was the Farndon Community Blockade between the 27th February and 30th March 2014 in response to a site originally permitted by Cheshire West and Chester Council on the 17th May 2010 and again on the 8th July 2013 for the drilling of an appraisal borehole for Coal Bed Methane exploration at King's Marsh, around three quarters of a mile (1.2km) due west of Farndon village at its closest point. The blockade was slow at first, but the numbers of people attending to protest outside the site entrance grew steadily each day until a permanent camp was finally set up on the 16th March – day 18 of the blockade. The tactics used were similar to other blockades with slow escorts of the lorries making deliveries being the main way of causing disruption, with around 50 to 60 people taking part at its peak – an extremely impressive turnout for what is an incredibly rural part of the country. Whilst modest in size and relatively short in duration the Farndon Community Blockade marked the UK's first organised resistance of a purely CBM site, with the one at Barton Moss being a combined CBM and shale well site.

Shortly before the Farndon Blockade ended, another camp was in the early stages of forming at another of Dart Energy's operations at Daneshill Energy Forest around 6.8 miles (11km) north-east of Worksop, near the village of Sutton-cum-Lound in Nottinghamshire. Nottinghamshire County Council granted the company permission on the 29th November 2012 for a 3 year extension to its previous permission to drill an appraisal CBM borehole that was due to expire. The community blockade lasted for a total of 47 days between the 26th March and 12th May 2014, and saw the most action taken against one of Dart's well sites. Equipment began to arrive and assemble on the site on the 26th March and so in response a small number of local residents began to set up a camp outside the entrance. It was not long before the protectors took part in non-violent, peaceful direct action. On the 27th March, one person managed to lock on to the top of the site entrance gates whilst two others locked on to a lorry attempting to make a delivery that had been stopped in the road outside. Banners were suspended from the gate and the vehicle, whilst another lorry transporting the drill rig was reportedly prevented from passing the railway level-crossing. As the blockade went on, more and more people began to arrive to take part and on the 31st March another successful complete block of the entrance was made by protectors wearing protective clothing and gas masks – to highlight the issue of ground, water and air contamination – who placed red and white cordon tape across the gates and prevented any access to the pad. With police presence minimal, subsequent temporary blocks of vehicles accessing the site were made on several occasions in the ensuing days. On day 20, 15th April, one protector succeeded in entering the site and scaling the drill rig, where he proceeded to 'occupy' it and unfurl a 'No Fracking' banner down its flank, stopping work for several hours. Nine days later, as the rig was in the process of being taken down, a Family Fun Day was held at the camp to celebrate the eventual end of the drilling and to thank those protectors who had been camping out since late March. The final lock-on of the blockade came on the 28th April, when a single protector attached himself by the wrists to the entrance gates, preventing any vehicular movements. The following day, a large police escort assisted the final removal of the drilling rig from the site. With almost all the on-site equipment having been removed and the pad abandoned for the time being, the camp also packed away, ending 47 continuous days of community blockades.

Returning to Cheshire, but remaining with Dart Energy, the company received a second permission for a site approximately one third of a mile (0.5km) outside the Chester suburb of Upton from Cheshire West and Chester Council on the 12th July 2010 and again on the 28th May 2013 for another Coal Bed Methane appraisal borehole, drilling down to around 5,577 feet (1,700m). Beginning on the 5th April 2014, the local communities reacted proactively to the impending threat placed upon them and managed to set up a community protection camp on the proposed site before any equipment could be moved on site. As at the time of writing, Dart have still made no public intention of beginning work at this particular site in the near future, but the Upton Community Blockade is still going strong. In addition to CBM exploration, on their website Dart Energy state that their Cheshire-based PEDLs also have potential for shale gas and that the company is keen to begin developing the potential of this unconventional energy source in the UK.

Heading now to the East Riding of Yorkshire, we have already discussed the Rathlin Energy projects at Crawberry Hill and West Newton. Two simultaneous community blockades were instigated on the 11th May 2014, following the standard method of delaying every delivery lorry and blocking the site whenever possible. Although the actual drilling occurred in the Summer of 2013, the two Community Protection Camps were established primarily in response to the mini fall-off testing that is feared to either be synonymous with an hydraulic fracturing operation – albeit one that is not intended to produce any oil or gas – or that it is a precursor to a fracking operation in the future. At Crawberry Hill, a permanent camp was erected outside the site entrance and included the construction of a wooden observation tower, affectionately known as the Crawberry Hill Castle, in order to provide a vantage point for watching over the activities on the site (i.e. ensuring planning regulations are being enforced). On the 2nd August, a second occupation of the pad itself was broken up by a combined police and bailiff eviction team, during which six people were arrested and, although the outer camp was not evicted, the 'Castle' was torn down. Nonetheless, the protectors were unperturbed and proceeded to re-build the camp and the Castle over the following few days.

A similar camp was established outside the West Newton site before being abandoned in November when Rathlin Energy pulled out of the site. Despite this, a Possession Order was pursued in the High Court by the landowner in early December in which the Judge ruled that the camp be cleared within 14 days and awarded costs of £10,000 against a single named defendant who had been residing at the camp to monitor the operations and the effects on the environment.

A number of allegations of incidents have been reported for the two Yorkshire sites, mostly with regards to the West Newton pad. On both the 11th and 23rd of May 2014, photos emerged apparently showing the protective perimeter ditch around the West Newton pad overflowing and flooding into the surrounding fields. The ditch is a common feature of well pads and is designed to trap any surface water runoff within a waterproof membrane to prevent any contamination of the ground. This was followed on the 29th May with photos and a video said to show a tank full of an unknown fluid leaking at the Crawberry Hill site. A couple of days later, back at the West Newton pad, another photo appears to show surface water once again flowing off the drill pad and onto the surroundings on the 1st June. This was followed on the 2nd July by large convoys of 64 vehicles entering the site, which was believed by the protectors to have been contrary to the information contained in the Traffic Management Report. Additional photos of apparent ditches full of water (including dead animals) at West Newton were published in early July. However, the most controversial incident reportedly occurred on the 12th August when an unexpected pressure build up between two layers of casing was detected, and speculation about the cause and extent quickly spread after specialist well inspection equipment was seen to be entering the site two days later. However, both Rathlin and the Health and Safety Executive maintain that the well has not been compromised. More recently, on the 23rd and 24th August, a contractor working at the Crawberry Hill site was photographed leaving the site with illegally covered number plates and the police were informed. On the second occurrence, the Community Protection Camp warned the police that the car was leaving the site and that they would block the road in both directions to stop the car if the police did not arrive, which they did after registering it as a high priority call. The driver was reportedly given a formal caution over the offences. The Crawberry Hill Community Protection Camp is ongoing at the time of writing.

Back to Balcombe...

Within a matter of months of their drilling operation in Summer 2013, Cuadrilla had submitted a new application to West Sussex County Council on 5th December 2013 for permission to return to the infamous Lower Stumble Wood site to conduct flow testing, including the flaring of gas since they had run out of time earlier in the year. Whereas the original planning application in 2010 received no objections or comments this later one attracted 889 objections to only 9 supports, no doubt thanks to the high profile protection camp and media interest. Despite this, West Sussex County Council voted almost unanimously on the 2nd May 2014 to allow Cuadrilla to return to Balcombe at a rather heated meeting that had to be suspended on two occasions at the behest of the committee chairman. Only one member of the Planning Committee voted to oppose the application, receiving great applause for doing so. Shortly afterwards, the No Fracking in Balcombe Society (NoFIBS) lodged an application with the High Court for a Judicial Review into the decision made by the Council. The proceedings began on the 6th November with the Barrister representing the residents, David Wolfe QC, stating that the Council's Planning Committee was given wrong advice by the officers when determining the application in May. His arguments centred on four main points. Regarding emissions, Mr Wolfe raised the fact that the Council were advised to include monitoring of sulphur dioxide for the proposed 46 feet (14m) high flare stack, but the committee were told that this was not a material consideration since the Environment Agency would oversee the pollution monitoring, despite the fact that the Environment Agency had already stated that it would not be monitoring for sulphur dioxide.

On the issue of well integrity, Mr Wolfe raised the concerns of the local residents over the condition of the old 1980s well drilled by Conoco at the same site. He argued that the officers told the Committee in May that the Health and Safety Executive had conducted in-depth investigations on the old well bore. However, Mr Wolfe continued by announcing that the Health and Safety Executive had been in discussion on this subject and revealed that there had not in fact been any investigations of the old well, and that the Council was advised that there had not been enough information contained in Cuadrilla's planning application for an assessment to be made. Mr Wolfe continued by saying that the Planning Committee had not been made aware of the correspondence. Other issues raised were the matter of previous breaches of planning conditions during the initial drilling operations in 2013 and the scale of objections to the application subject of the Judicial Review. In the afternoon, the Council's Barrister countered the claims by Mr Wolfe and defended the actions of the Council, describing how, in his opinion, all the rules and regulations were followed correctly. The Review concluded the following day, 7th November, with a summary by Mr Wolfe of the reasons for the case against West Sussex County Council and asked that the judge quashed the permission granted in May. The Judge's decision finally arrived on the 5th December in which he ruled against the Residents' Association and in favour of the Council. Whilst expressing sympathy with the claimants, he branded the case as "misconceived [and] without merit" and said that the argument that the Planning Committee were mislead could not be sustained on the facts. Costs of £10,000 were awarded against the Frack Free Balcombe Residents' Association, though the green energy company, Ecotricity, offered to pay the fees.

However, at the beginning of September 2014, a Balcombe resident scored an impressive victory against the head of American oil giant Breitling Energy Corporation, Chris Faulkner, the so-called 'Frack Master'. In February, Mr Faulkner had taken out a full page advertisement in the Sunday Telegraph newspaper in which he made a number of claims about shale gas and UK energy security. A complaint was made to the Advertising Standards Authority (ASA) by a Balcombe resident, listing six claims that were misleading or unsubstantiated. The ASA investigated the complaint and finally announced that they upheld all six complaints which included the use of emotive language that was likely to be interpreted as a statement of fact; the amount of recoverable shale gas is unknown, though estimates suggest there is only nine years worth of recoverable gas and not the decades worth as indicated in the advert; it is not possible to calculate likely tax revenues from shale exploitation; the UK had not suffered any interruption in gas supply as a result of disputes between Russia and Ukraine; there is insufficient evidence to support the claim that shale gas would reduce energy prices; and finally that the claim shale gas would reduce greenhouse gas emissions is uncertain and misleading.

...Lancashire in the Limelight

2014 has seen a new wave of applications by Cuadrilla for their operations on the Fylde Peninsula in Lancashire. As at the time of writing all are still to be determined. The first was submitted to the Council on the 27th March to seek permission for the retention of the Becconsall (Banks) well pad for a further three years to allow pressure monitoring of the Bowland Shale reservoir. If permitted, the application would see the existing drilled well perforated and then temporarily plugged in order to create a pressurised test zone for monitoring of the shale layers. No new fluids would be injected, but the existing fluid within the test section would be cycled inside the well. Cuadrilla state that, if permitted, the works would not begin until 2015 and would avoid taking place during the bird nesting season that had previously caused the company to abandon its plans for the Annas Road site. More than 244 comments have been logged for this application. This was followed on the 19th May for a near identical application for the Grange Hill site near the village of Singleton. On this occasion, Cuadrilla are also seeking the permission to install seismic monitoring equipment above and below the test section of the well. Over 255 comments have been submitted to this application. Then, on the 29th July, the company applied for permission to once again extend the period of time allowed to restore its Preese Hall site until the 30th April 2015. This is the third time that the company has sought extensions for the restoration of the site to its previous use – the first was submitted on the 14th June 2011 to extend the period until the 31st December 2013; the second was submitted on the 19th December 2013 to extend permission until the 31st July 2014. Both were granted on the 23rd January 2013 and 3rd March 2014 respectively.

Although it can be inferred from the first two applications mentioned above – Becconsall and Grange Hill – that shale gas extraction could be expected within the coming few years, of more immediate concern to Lancashire in particular, and to the wider anti-fracking community, are four simultaneous applications for two separate sites that expressly seek permission for the UK's first large-scale hydraulic fracturing operation. These are for up to a total of 8 production wells, totaling 9.9 miles (16km) in combined length, an additional 12 water and gas monitoring boreholes and 180 seismic monitoring sensors over a combined area of up to 38.6 square miles (100km^2). The first proposed site, just north of Preston New Road at the village of Little Plumpton lies 2.4 miles (3.9km) due south of the 2011 Preese Hall fracking site, whereas the second is at the village of Roseacre around 4 miles (6.4km) due east of Preese Hall. Scoping Opinions for both sites were submitted to Lancashire County Council on the 4th February 2014 before the full applications were submitted. Each site had two separate, but linked, applications, each dealing with the two main operations – the drilling and fracking, and the installation of seismic monitoring arrays. Each pair of applications are identical for both the proposed sites. The seismic monitoring applications – submitted on the 2nd June (Preston New Road) and 17th June (Roseacre) – seek permission for the drilling of three pairs of boreholes (six total per site) around the well pad to a depth of

65 to 98 feet (20m to 30m) for the monitoring of groundwater and gas migration in addition to the installation of 80 underground and 10 surface seismic arrays within a 2.4 mile (4km) radius of each pad. The 80 underground sensors would be buried to a depth of around 2.6 feet (0.8m) and would be almost invisible on the surface except for a small fenced off enclosure and a square-shaped cap at ground level to cover the holes. The surface sites would also include a small electrical equipment cabinet. Neither application has been determined by Lancashire County Council as of the time of writing.

The accompanying applications for the main drilling application are again identical in their proposal for both the sites. For each pad, Cuadrilla are wanting to drill up to four wells and openly state they intend to hydraulically fracture and test each one. On each pad the first well is proposed to be drilled vertically to 11,483 feet (3,500m) deep into the Bowland Shale and a lateral well splitting off the vertical at between 4,921 and 11,483 feet (1,500m to 3,500m), extending up to a distance of 1.2 miles (2km). Once this well has been fully fracked, wells 2 to 4 would then be drilled as lateral wells at varying depths and extend up to 1.2 miles (2km) out from the pad. Cuadrilla propose to conduct what they term as 'mini-fracks' before the main fracturing operation within the wells in order to calibrate the seismic and other monitoring equipment. Mini-fracks could also be undertaken at various stages during the rest of the operations for similar purposes. The company describes a 'mini-frack' as injecting fracturing fluid without the **Propant**, in this case silica sand – a well known cause of the respiratory condition silicosis – under pressure to fracture the rock, but allow the fractures to close again so as to limit the amount of oil or gas entering the wellbore and be wasted. In the main operation, Cuadrilla state that the hydraulic fracturing would "be performed over 30 to 45 stages per well at intervals of 30 to 50m [98 to 164 feet] per stage." This means that in total 180 separate hydraulic fracturing stages could be carried out per pad, or 360 in total across 11 miles (18km) of boreholes across the two proposed sites! As a 'contingency' they also state that if necessary, each fracture stage could be pre-stimulated with a 10% dilution of hydrochloric acid to acid-frack the shale in order to soften it and allow it to fracture more easily under pressure. For the Preston New Road site, the application was submitted on the 2nd June 2014 and over 1,250 comments have been uploaded to the Council's website already, though thousands more are known to have been submitted. The Roseacre application was submitted on the 17th June 2014 and has over 550 comments logged with the Council. As with all of Cuadrilla's Lancashire applications submitted in 2014, neither of these have been determined as of the time of writing. Regarding the Preston New Road proposal, the air safety organisation, NATS, initially registered a formal objection on grounds that the site would "reduce the probability of detection of real aircraft at low elevations as well as the generation of false tracks on the RSS St Annes SSR". However, NATS withdrew this objection on the 14th August 2014.

Contemporary with the new fracking applications, the Government's policies on shale gas in the period leading up to the 2015 General Election were also formally announced during the annual Queen's Speech in June 2014. This included giving drilling companies the ability to drill under any land without having to gain permission to do so at depths of 900 feet (300m) or more. Whilst Parliament were assembled in the House of Lords to hear the Queen reading the programme for the next sitting of the Commons, a group of activists from Greenpeace dressed as drill site workers fenced off Prime Minister David Cameron's home in Oxfordshire ready for it to be fracked. They also posted a giant novelty cheque for £50 through his letterbox – said to be the equivalent compensation landowners and homeowners would receive for drilling under their property – in a mock gesture at what is termed as a bribe to local communities to accept fracking.

Balcombe to Belcoo

On the 1st April 2011, the Northern Irish Department of Enterprise, Trade and Investment (DETI) granted Tamboran Resources a petroleum licence for a large area on the Irish border, in the south-west of the country. The company describes itself as "a private international exploration company focused on finding and developing unconventional oil and gas resources in multiple basins around the world." It currently holds licences for Northern Ireland and Australia whilst expecting to receive another for the Republic of Ireland in 2015. In the licence for the area including County Fermanagh, the DETI set out an agreed schedule for the exploration and drilling programme that the company is expected to fulfill in order to retain the licence area, of which clause xi states "Drill exploration well to test Benbulben and Bundoran Shale Formations gas shale play, including coring, fracturing and testing programme" within 4-5 years of gaining the licence (i.e. 2014 to 2015). Furthermore, clause xvi also states "Drill 2nd exploration well to test Benbulben and Bundoran Shale Formations gas shale play, including fracturing, multiple horizontal legs, flow testing and seismic monitoring". Despite this, Tamboran's website says regarding its Northern Irish licence that they expect "to initiate drilling of a scientific borehole in Co. Fermanagh in 2014, as part of fulfilling the terms of its exploration licence…The purpose of this drilling operation will be to collect rock samples and no hydraulic fracturing will take place".

A fracking awareness campaign started up in late 2011 in response to the news that Fermanagh could become the frontline of Irish shale gas exploration and production and on the 21st July 2014 it was announced that in the early hours of the morning the company had moved into a quarry around 2 miles (3.5km) east of the small village of Belcoo on the Irish border. Community resistance immediately jumped into action and by that same evening, over 300 local people had amassed outside the entrance gates. The first ever Irish anti-fracking protest was beginning to occur and by the 27th July, the number of people taking part in the blockade had risen by another 100, around the same number as took part in the Balcombe Blockade during the Summer of 2013. Tamboran had utilised an anomaly in the Northern Irish planning system that meant the company could move onto the site and assemble its equipment without the need to gain permission to do so. In response, on the 30th July, Fermanagh and Omagh District Council passed a motion to oppose fracking and called upon the Northern Irish Assembly to take action on the issue. In the Republic of Ireland, a government-imposed moratorium on the process has been in effect whilst a wide ranging environmental investigation into potential impacts is taking place. Responding to the mounting pressure from residents and authorities alike, the Northern Irish Environment Minister refused to allow Tamboran to drill on the site under permitted development rights, stating that an Environmental Impact Assessment is required and therefore the company would need to apply for planning permission.

On the 26th July, Tamboran obtained an injunction through the High Court in response to the blockade which makes it unlawful for unauthorised people to enter or occupy the site or otherwise prevent access into and out of the site. Nonetheless, the protest and protection camp has continued on the road outside, and on the 3rd August, a large number of farmers from the region took part in a mass anti-fracking tractor protest in which 140 tractors drove from the Belcoo site to the town of Enniskillen to raise awareness of the dangers of hydraulic fracturing and the impact it could have upon their businesses. The Community Blockade, though hindered by the injunction, continues to maintain its presence outside the site.

Frack-Attack – Round Two

With news of Cuadrilla's latest applications to bring mass hydraulic fracturing operations to rural Lancashire, the seventh community protection camp and blockade was initiated at the Preston New Road proposed site on the 7th August 2014. Being only the second blockade to operate in advance of works beginning on site, and the first to take place before any applications have been granted, a number of local residents, mostly comprised of mothers, grandmothers and children, set up the camp on the proposed site in order to prevent Cuadrilla or their contractors to gain access and oppose the multi-well application. Later that morning, another contingent of the newly formed Frack Free Lancashire coalition handed in a record breaking 14,000 objections to the application that is currently under consideration. With the UK planning system being seen as largely ignoring local communities and giving preferential treatment to developers, non-violent, peaceful direct action such as this is seen as the only way that the public can attempt to make their voices heard, particularly if access to elected representatives is perceived to be difficult or fruitless. Despite Government promises of Localism, more and more people are waking up to the faults and peculiarities of the planning system that tends to be more exclusive, rather than inclusive, of the general public. Seven days into the blockade, and with numbers of participants growing in opposition to Cuadrilla's plans, security cameras were erected in a nearby field, allegedly by or on the behest of the company, to monitor the camp and any activity taking place.

On the 23rd August 2014, the Blackpool Gazette announced that 10 landowners around the Preston New Road site, supported by Cuadrilla, are seeking a High Court injunction similar to that gained by Tamboran in Northern Ireland to evict the protectors off the site and "prevent future occupations at sites on and around the planned drill sites, at Preston New Road and Roseacre Wood". The Lancashire County branch of the National Farmers Union was also reported to be supporting the court action that occurred on the 28th August. Responding to the announcement, a representative of No Dash for Gas said "This legal action is a last gasp attempt by a dying industry that has failed to gain the required social license and refuses to know when it has been beaten". The Court ruled in favour of Cuadrilla with the support of the defendants after making significant reductions to the terms on the injunctions and agreed for a call by the defendants that the temporary injunction should be reheard on the 8th October to allow for a full assessment and objections to be raised.

In the meantime, Fylde Borough Council held a meeting on the 18th September 2014 to determine its recommendation to Lancashire County Council with regards to the Preston New Road and Roseacre sites. With over forty members of the public who addressed the Committee to support the Officer recommendation to refuse, the Borough Council eventually agreed to request that the County Council should vote against both sites. The final decision by Lancashire County Council is yet to be made at the time of writing.

Reclaim the Power
In Summer 2013, the last minute relocation of the No Dash for Gas 'Reclaim the Power' environment camp forced a week long suspension of operations at Cuadrilla's Balcombe drilling site. In 2014, the camp took place at the Preston New Road blockade between the 14th and 20th August, with around 1,000 activists from across the UK taking part. The first event came on the 17th August with a mass protest march between Blackpool's two piers attended by hundreds of local people and members of the protection camp.

As occurred at Balcombe in 2013, the first Monday of the camp, the 18th August, saw a national day of direct action against the unconventional oil and gas industry and financial partners. The first of the day's thirteen separate actions occurred at 6am at Swansea University's Bay Campus. The University is the focus of a national Energy Safety Research Institute, funded by Government and industry sources, which conducts research into, amongst other things, the extraction of fossil fuels, including research into hydraulic fracturing and other unconventional methods. A number of Swansea residents, students from the University and graduates dressed in 'mad scientist' outfits and occupied the site, stopping the construction work for much of the day. Two people had locked-on in various parts of the site, whilst the rest gained entry to the building, occupying it and unfurling a 'No Fracking' banner from the roof. Outside the campus, a second large banner reading 'Dim Ffracio' – Welsh for 'No Fracking' – was also displayed prominently. At the same time, families in Lancashire planted 88 large, biodegradable models of Radium atoms around the town of Lytham St. Annes in a temporary community arts installation to highlight the dangers of fracking, in particular the radioactive material that is brought to the surface through the injection of fluids deep into the underground geology.

Meanwhile, in London, a group of activists 'occupied' the offices of the Department for the Environment, Food and Rural Affairs (DEFRA) in response to a report by the Government Department released the previous week on the potential impacts of shale gas exploration on rural communities. The report was only 13 pages in length but contained 63 redactions, obscuring almost all of its content. Eight sections had been deleted from the executive summary, four sections on economic impacts, four sections on social impacts, 17 sections on local service impacts and three sections specifically looking at the impact on house prices near drilling sites. In late August, the Daily Mail ran an article detailing how many homeowners and estate agents in England are reporting impacts upon property values in areas earmarked for oil and gas exploration, including the story of one person near Cuadrilla's latest proposed sites in Lancashire whose £725,000 home had been newly revalued at just £190,000 (a 74% decrease) even though no permission has yet been granted for the site. Brighton Pavilion MP, Caroline Lucas – arrested at the Balcombe protest in 2013 – criticised the redactions and accused the Government of purposefully withholding information on the dangers of fracking, demanding that the public be given the chance to read the report in full. At the DEFRA offices, three activists glued themselves to the main doors whilst others blocked the entrance in a three-person lock-on involving reinforced arm tubes designed to prevent police or security officials from forcibly breaking them apart. Another member managed to scale the building to unfurl a large banner stating "What's to hide DEFRA? Don't frack with our future". Many of the participants also had their mouths covered with black tape to highlight the amount of data being hidden in the report. Staying in London, the fourth action was a blockade of the IGas headquarters in Westminster by 15 protectors from many of the former and current community blockades around the country. IGas were responsible for the operations at Barton Moss during the Winter and Spring of 2013 to 2014 and are said to be the largest unconventional oil and gas company in the UK.

At the Rathlin Energy site at Crawberry Hill, protectors succeeded in completely blockading the site with a number of lock-ons at the outer gate and the displaying of a large banner on the security fence, whilst others presented the site with a mock Safety Report that had the word 'FAIL' branded across it. Several reports of aggression by site security personnel were made by the protectors. Another group glued themselves to the doors of Total Environment Technology who are a contractor to the Crawberry Hill and West Newton drilling sites.

A short distance from the Preston New Road proposed site, 11 members of the Reclaim the Power camp proceeded to occupy Cuadrilla's northern offices and perform lock ons to close down the building for the day, whilst a large crowd of the anti-fracking movement gathered outside to protest publically, not least because the Cuadrilla offices are opposite the North and West Chamber of Commerce offices, who have been supportive of the company's operations. In contrast, the owner of a large plant nursery business in close proximity to Preston New Road voiced his support for the Reclaim the Power camp and is highly concerned about the future of his business if Lancashire County Council grant permission for the multiple-well site. In Salford, campaigners attached a huge banner on a bridge across the Manchester Ship Canal at Salford Media City displaying the fact that an extremely huge quantity of waste water from Cuadrilla's Preese Hall site was disposed of into the Manchester Ship Canal, after treatment at the waste water works, the previous year. A similar banner was displayed at Blackpool College who stand accused of accepting donations from Cuadrilla. In the afternoon, 13 activists staged a 'die-in' event at a Blackpool branch of HSBC, who are accused of being the UK's single largest investor in fossil fuels and providing corporate banking services for Cuadrilla. Displaying banners

Fig.26: 'Cuadzilla' has become a familiar sight at fracking campaigns in the North-East, such as here at the IGas rig at Ellesmere Port, 14th December 2014

highlighting the bank's investments and the dangers of hydraulic fracturing, the activist spread themselves across the ground inside and outside the bank in an act of passive resistance in which they allow their bodies to go limp and play dead to make it difficult for security personnel to remove them from the premises. In Manchester, the offices of Political Planning Services, Cuadrilla's public relations contractors, were targeted by a small number of activists dressed in toxic hazard suits who engaged members of the public and other businesses sharing the same building on the subject of fossil fuels, climate change and renewable technology. The PPS Group were subject to a Channel 4 'Dispatches' and an Evening Standard investigation in 2007 during which they accused the company of an array of underhand tactics, including forging letters of support, in gaining permissions for their clients. In Blackpool, members of the Reclaim the Power Camp visited the homes of two local councillors with alleged interests in supporting fracking and Cuadrilla, and it is reported that residents of the same street joined in to support the action. In nearby Preston, another group from the camp posed as employees of a made-up fracking insurance company 'Fraxtons' and tried to 'convince' passers by to take out insurance cover with them as the only insurance company offering cover for fracking regions whilst actually highlighting the issues of house prices and home insurance that are already being reported by homeowners and estate agents across the UK.

With the national day of action over, the final event came on the 20th August with a cycle ride from the Preston New Road site to Hull, following part of the route of the 2014 Tour de France course. The cyclists – known as the 'Tour de Frac' – stopped at various points to chat with locals about fracking and to collect more letters of objection to deliver to Lancashire County Council. Ending up in Hull on the afternoon of the 23rd August, the riders joined a protest march through the city that ended with a mass rally to listen to a variety of speakers on the subject.

Rathlin Warned

Throughout mid to late September 2014, residents living around Rathlin Energy's West Newton site in East Yorkshire reported recurring odorous emissions from the well pad. This resulted in a wave of complaints to the Environment Agency alleging bouts of illness, nausea and stinging eyes and including reports that audible warning alarms could be heard coming from the site at times when the odour was at its strongest. Despite regular flaring of gas, the reports continued claiming that the odour and symptoms return shortly after the flaring ceases. On the 19th September, the Environment Agency warned Rathlin that they must stop the odour. This was ten days after the first contact with the company over the matter. New procedures were reported by the company to have been implemented to reduce and remove the odour – believed to be emissions of un-burnt natural gas from the flare.

Celtique's First Well

News broke on the morning of the 15th September 2014 that prepatory work had began at the Broadford Bridge site near the village of Billingshurst in West Sussex, a few miles east of the proposed Wisborough Green site. Lacking a suitable area for a protection camp, a small group of residents decided to hold a weekly awareness-raising vigil a short distance from the site, at a notoriously dangerous crossroads that all the works traffic would have to use to access and leave the site. Although Celtique claim that the well was only targeting conventional formations their investment partner, Magellan Petroleum, have stated that the well will be logging and coring shale in the Kimmeridge Clay and Liassic strata. They also make an explicit reference to the Broadford Bridge site being an unconventional exploration well in an investor presentation earlier in the year, and the residents were keen to make the local inhabitants aware of this fact – many of whom were not even aware that an oil drilling site was being developed such a short distance from their homes and a small pre-school.

Beginning in early October, a weekly four-hour roadside vigil continues to be held every Saturday beside the crossroads, with a 'Frack Free Zone' banner and leaflets to hand out to passing pedestrians. One local campaigner taking part in vigils – a local resident of more than twenty years – said "I knew that these crossroads were dangerous, but I didn't realise just how dangerous. There have been so many near-misses with people not paying attention, using their phones or simply being impatient. I dread to think what it will be like once the HGVs from the drilling site start using these roads."

Fig.27: A member of The Warrior's Call leaves a votive offering and poem at Broadford Bridge in memory of a large oak tree that was felled to make way for the drill site.

Fig.28: The Adversane Crossroads will have to be navigated by traffic serving the Broadford Bridge well site. The verge on the left has been the scene of a weekly vigil

A well-attended public meeting was held on the subject of the drilling operations at Billingshurst Community Centre at the end of the month, at which guests from

Balcombe, Fernhurst and Wisborough Green told of their experiences of dealing with oil and gas operations, and in the case of the latter two villages, their experiences of Celtique Energie. At the meeting, it was also agreed that a convoy of local residents should be organised to make a visual demonstration of public feeling as well as highlighting the traffic impact of the site. On the 22nd November, around 20 vehicles decorated with balloons and banners drove from the outskirts of Pulborough, north to Billingshurst and then south again to head past the drilling site whilst the usual roadside vigil at the crossroads was manned by supporters. The event was well received by the community, and local campaigners saw an increase in interest expressed by people wanting to find out more about the site. A few days after the event, news was released that 50% of residents in the nearby village of Pulborough were moderately or strongly opposed to fracking compared to only 21% support, in a survey conducted by Pulborough Parish Council's Neighbourhood Plan Steering Group during August. The survey also revealed that 80% of residents supported local renewable energy development, with 44% favouring the idea of land being set aside for solar arrays.

Fig.29: *The protest convoy passes the village of Billingshurst, 22nd November 2014*

At the time of writing, the rig has yet to arrive at Broadford Bridge due to construction of the well pad being delayed because of the ground being too wet, according to an employee from the Environment Agency. This has, however, only increased the concern by those opposed to the operations about the risk of contamination to groundwater if more water is present than was previously expected. Shortly before going to publication, campaigners local to the site received copies of Magellan's quarterly financial returns to the American Securities and Exchange Commission from March, June, September and November 2014. These documents revealed that the drilling at Broadford Bridge is expected to commence in the second quarter of 2015 following a "planned delay for weather related concerns" and that "Pending the results of this well, the Company may participate in a second exploratory well...in fiscal year 2016". The quarterly return for March 2014 also further confirmed fears that the shale-bearing layers will be explored:

> "A complete suite of logs and cores is planned to be collected from the Kimmeridge Clay and Liassic formations, which we believe will provide technical data, including thickness, oil maturity, formation pressure, and rock brittleness, to be able to assess the potential for unconventional development of these formations and in turn possibly attract partners to continue the development of these licenses."

The total cost of the operation has been estimated at $10 million, based upon Magellan's stated 50% share of the cost being $5 million, of which they have already invested nearly $2 million.

The future of another appraisal well co-owned by Magellan drilled at Markwell's Wood near the village of Rowlands Castle on the West Sussex-Hampshire border was also revealed in the June report. This well had been drilled by Northern Petroleum between the 21st November and 26th December 2010 during which a small, conventional reservoir was found. The company commissioned Schlumberger to investigate the finding of the drilling, which revealed that "the area is probably immature for oil or gas generation and therefore unlikely to have unconventional shale oil or gas". With this news, the September report stated that Magellan and their PEDL partners, Egdon Resources and Northern Petroleum, had decided to dispose of the well by attempting to sell off or farm out the license to a third party, or failing that, "If the Company and its partners are unable to sell or farmout [the license], the Company may face a plugging and abandonment liability of approximately $378 thousand net to its interest".

The Peoples' Climate March

On the 21st September 2014, the largest Climate March in history took place across the globe in advance of a UN Climate Summit called by Secretary General Ban Ki Moon for the 23rd September. On the day, more than half a million people in 162 countries took part in 2,646

Figs.30 and 31: Before and after at the Broadford Bridge site

separate marches to call on the Summit to agree to urgent and immediate action to combat climate change and move away from fossil fuels. A host of A-List celebrities lent their support to many of the marches and Ban Ki Moon himself also headed the largest march in New York City. The numbers of participants were four times larger than initial estimates expected at many of the marches, including in London where more than 40,000 people took to the streets.

October 2014 – Two More Camps

In early October 2014, a small group of campaigners began to hold a weekly roadside vigil near the Horse Hill drilling site in Surrey to highlight to passing motorists what was happening at the site. The reception was better than expected and so a picnic at the site was organized by a number of locals to help further the local awareness raising. Soon after, a permanent roadside camp outside the site entrance was established and has continued to be manned by a small number of people since. On the 22nd October, two campaigners staged a lock-on in front of the gates, one of whom was later arrested. A few weeks later, on the weekend of the 7th to 10th November, a wave of activity occurred in response to the dismantling of the rig and other equipment. On the Saturday another two-person lock-on in front of the gates prevented the rig from leaving for more than 48 hours whilst on the 10th the protectors were permitted, despite some threats of arrest, to slow march the remaining vehicles from the site to the main road. Meanwhile two campaigners managed to scale two tankers as they were leaving the site.

As with the Broadford Bridge site, Magellan Petroleum also had a large stake in the operations at Horse Hill and had previously announced that they were to fund the coring and logging of the shale layers whilst the operator was adamant that the drilling was purely conventional. Shortly after the drilling began, Magellan made another announcement that it had in fact decided not to drill the shale layers.

Later in the week following the demobilization of the site, the Australian CEO of Horse Hill Developments Ltd, David Lenigas, announced in an interview for the online version of the Daily Mail some of the findings of the drilling operation. He stated that the company were looking to begin production form the well during 2015, which could include the drilling of

Figs.32 and 33: The Horse Hill drill site was highly visible in the landscape, including from some distance away

a second, perhaps horizontal, well to assess another conventional layer. Another announcement that caused much concern to the anti-fracking community was of an 'unexpected' find of hydrocarbons in the Kimmeridge Clay (shale) layers, despite the previous announcement by Magellan that the coring and logging of the shales had been called off. Referring to this find, Mr Lenigas – who had previously ruled out fracking at Horse Hill – said "We are quietly optimistic that we will have some additional oils to the recoverable potential for Horse Hill", raising fears and adding to the suspicion that the next phase of operations being planned for 2015 could include the use of hydraulic fracturing.

Investors were also left reeling after the results had estimated that the potential oil reserves at Horse Hill were grossly reduced from pre-drilling estimates. Magellan Petroleum's interests were also hit, with their share prices falling from more than $1.40 on the day of Mr Lenigas' interview to just $0.92 within two weeks – a fall of over 34% – though how much is attributable to the Horse Hill findings and how much to other external factors cannot be certain. At the time of writing, Magellan's share prices have fallen further still to $0.85, a 50% decline since November 2014.

News broke on the 27th October 2014 that a new camp had also sprung up at Burton-on-the-Wolds in Leicestershire to help inform the local residents about the activities of Egdon Resources who had began to drill at nearby Horse Leys Farm. Whilst the camp claims to have received much support from the local and wider communities alike, an article in the Leicester Mercury carried a statement from the Vice-Chairman of Burton Parish Council attacking the camp:

Fig.34: Horse Hill, 4th October 2014

> "It's a pity [the campaigners] haven't got anything better to do. They must not have jobs."

Aside from the widespread misconception that campaigners are unemployed and/or receiving state handouts, the above statement was refuted by the camp:

> "In response to that comment I will say there is nothing better to do then try and safeguard our young from the greed- coloured judgement of men. History is full of such harm."

Borras Borehole

The beginning of October also saw a previous decision by Wrexham County Borough Council, in Northeast Wales, to refuse GP Energy (a subsidiary of Dart Energy) permission to drill an exploratory borehole for Coal Bed Methane appraisal near Borras, overturned by the Planning Inspector appointed by the Welsh Assembly Ministers.

The application was submitted on the 20th September 2013 and indicated that a single vertical borehole would be drilled to a depth of around 3,840 feet (1,170m) in order to remove a sample of the geological formations. By the time the application was considered by the Council, 38 objections from local residents (including the local Assembly Member and the Member of Parliament) had been submitted in addition to a 1,649 signature petition. The Concern Communities of Falkirk group also submitted a document detailing the issues they have faced regarding similar exploratory drilling applications, and the ongoing public enquiry. Despite this, the Council were given the recommendation to approve the application by the Case Officer. On the 31st March 2014, the Planning Committee voted to refuse permission, citing it was environmentally unsustainable, represented industrial development in a countryside location and that the geological survey lacked sufficient information.

An appeal was lodged and was subsequently approved on the 8th October 2014, to the disappointment of many local residents and the local Plaid Cymru Assembly Member, Llyr Llyr Gruffydd. A local meeting was immediately arranged for later in the month to discuss plans on how to tackle the drilling operation and on the 14th October the Plaid Cymru group at Wrexham County Borough Council called on all 52 councillors to support their motion to make Wrexham a Frack Free Zone.

A Community Protection Camp was quickly established on the site on the 17th October and soon received a great deal of support from the local communities. However in mid-November a court summons was placed on the camp and the judge ruled in favour of evicting the camp at a hearing on the 20th. The protectors were nonetheless defiant and vowed to remain on the site until fracking was banned, saying in a press release "the time for debate has passed. We cannot wait for the irreversible damage that unconventional gas extraction will cause." More than 2,200 people have supported the Protection Camp from across the UK, with many travelling to the site with gifts of food and other supplies.

Further controversy over the site came on the 26th November, when local residents and campaigners warned that the drilling would desecrate the graves of 255 miners who were killed in the 1934 Gresford Colliery disaster and whose bodies still remain trapped in the mines. Surviving relatives and friends of the victims also branded the proposed drilling as "disgusting" and an insult to the memories of those who were killed. One person, whose grandfather worked in an adjacent mine, said "the gas will be extracted from the lungs of dead miners for profit. This cannot be allowed to happen." The daughter of one of the victims also spoke out against the drilling, stating "it would be very insensitive to run a business close to all these graves. We want a five-mile exclusion zone for any work close to the graves." Another campaigner also highlighted the fact that a previous CEO of Dart Energy had supported drilling a short distance from the Auschwitz Concentration Camp in Poland in 2012. Despite this the police and bailiffs entered the camp on the 27th November, arresting a legal observer and proceeded to demolish the temporary structures accommodating the protectors. Dart Energy dispute the controversy over the Gresford Colliery, claiming that the well will be around 3,000 feet (1km) away from the grave site.

IGas Eyes on Sussex

On the 30th October, news broke of two proposed drilling sites in the South Downs National Park by IGas. Speaking on behalf of the company at a public meeting arranged by Graffham Parish Council, a land agent made reference to plans to drill a well at Baxter's Copse near the West Sussex village (the site of an uneconomical and abandoned exploration well drilled in 1983) and for a second well at West Dean, near Goodwood. Although stating that IGas had no plans to frack at either well, when questioned further on this, the agent said "we are an oil company, not a gas company, despite the name. If we found gas, rather than oil, we would have to go back to the planning process and start all over again." Furthermore, when questioned on local job creation, a common benefit claimed by many oil and gas companies, the response was "there won't be any jobs really." One resident also asked for details of the PEDL for the Graffham area, to which the rely was reportedly "I've never heard of PEDL licenses."

IGas is the company behind the drilling at Barton Moss and several other Coal Bed Methane sites in England and Scotland.

Fig.35: The IGas rig at Barton Moss, January 2014

News from West Newton

With the operations at Rathlin Energy's controversial exploratory site at West Newton in East Yorkshire completed on the 5th November, the company outlined its proposals for a third well nearby on the 12th. The new proposed site, closer to residential property than the other two (West Newton and Crawberry Hill), would expect to drill down to 6,890 feet (2,100m) and avoid the Bowland Shale. The news first came to air in a letter sent to 9,000 households in the Holderness area at the end of October, which invited residents to a public exhibition held on the 8th November. The event was restricted to residents in the targeted region and attendance had to be pre-booked, with proof of identity having to be shown on the day.

The Chairman of Rathlin Energy, David Montagu-Smith has

also been the subject of controversy by some members of the anti-fracking community over some of his other activities. Mr Montagu-Smith is also the Chairman of the West Northamptonshire District Committee of the Campaign to Protect Rural England (CPRE) and it has been revealed in the minutes of a meeting of Desborough Town Council held on the 20th February 2014 (at which he gave a presentation on fracking) that he worked on the development of CPRE's National Fracking Policy Guidance. The opening line of these guidance notes (published in November 2013) states:

> "Based on the information we have at present, the CPRE does not oppose the exploitation of shale gas in principle provided it meets certain conditions."

In response to the criticism in an article written by a member of the Kent branch of CPRE, Rathlin Energy issued a statement refuting the claims of irony and a conflict of interest and that "[Montagu-Smith] is quite certain that the countryside has nothing to fear from the work of the oil industry, provided all its operations are carried out in full compliance with all the many, detailed requirements…and are maintained to the highest standards of the industry…We are not hydraulically fracturing the wells we have drilled at both of our current sites in East Yorkshire and we have no intention to do so."

A Centre for Fracking Excellence?

November also saw the announcement by the Energy Minister, Matthew Hancock, of plans to create a UKIP-style 'soverign wealth fund' for the northwest of England, claiming in his speech on the 12th November that fracking "will transform the North." Mr Hancock also broke the news that the Blackpool and The Fylde College is to become a new 'National College for Onshore Oil & Gas', with offshoots in Chester, Portsmouth, Redcar and Strathclyde. Responding to this, Blackpool campaigner, Tina Rothery, said:

> "As a mother and local resident I am fuming. Our bloody children now! Bad enough they buy Councillors, lie to and pay off landowners, mislead residents and railroad campaigners, but this is obscene. These students being led into careers that are short-term, dangerous to their health, will likely involve them in costly litigation at some stage and make them accessories to this vile business."

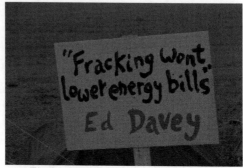

Fig.36: Despite several high-ranking politicians and industry leaders stating that fracking would not reduce energy bills, many pro-shale advocates still cite lower bills as a benefit

The news came on the same day that the UK Energy Research Centre warned Ministers over their promises of cheaper bills and energy security, dismissing them as "hype" and "lacking in evidence".

More Councils Turn Frack-Free

On the 11th November, Fingal County Council in Ireland passed a motion raised by Labour Councillor Brian McDonagh which stated "That this council is opposed to Fracking in Fingal or anywhere on the Island of Ireland." Just two days later a shock decision by the Conservative-led Trafford Council passed a similar anti-fracking motion stating that it will oppose fracking "until such time as it can be proved to be safe", bringing the total number of official and unofficial bans in the UK and Ireland at the time of writing to 27. Wirral Council was the first to declare concern about fracking back in July 2011, followed by the Conservative-led

East Sussex County Council in March 2012. However, a later petition calling for East Sussex to be 'upgraded' to Frack-Free status was refused in October 2014. The first Council to declare itself a Frack Free Zone (as opposed to simply expressing concern or opposition to the technology) was the Green-led Brighton and Hove City Council on the 24th January 2013. A number of other administrations across the UK joined the list of those whom have passed resolutions of concern and opposition in November 2013, followed by another wave in January 2014.

Other Councils have refused to take any stance of the matter. West Sussex County Council, who have had to deal with applications for exploratory drilling at Balcombe, Broadford Bridge, and Wisborough Green, debated a 3,000-signature petition by West Sussex residents to impose a temporary moratorium on fracking until it is either proven to be safe or other, renewable technology makes fracking for oil and gas obsolete on the 17th October. Following a short debate, a vote was taken on the petitioners' request. Of the 61 members present at the vote, 52 (41 Conservative, 8 UKIP, 2 Liberal Democrats and 1 Independent) voted against the petitioners request, six voted in favour (5 Labour, 1 Liberal Democrat) and three abstained (2 Liberal Democrats, 1 UKIP). The Cabinet Member for Highways, Transport and Fracking then proposed his own, alternative, motion which called for the Council to then vote on not to take a vote on fracking. This second motion was passed.

Another Council to oppose an anti-fracking initiative was Canterbury City Council, which voted 27 to 7 against the motion, with five abstentions on the 27th November. The text, proposed by Liberal Democrat Councillor, Mike Sole, stated:

> "This Council is opposed in principle to hydraulic fracturing and all forms of unconventional gas and oil

exploration and development in the Canterbury District, particularly on its own land. As landowner we will not allow such exploration and development on our land, although we would consider all planning applications on the planning merits of each. We support renewable energy sources such as solar, tide and wind."

The negative result of this vote came as news leaked out to the public of plans for drilling in the Tilmanstone area of East Kent. Coastal Oil and Gas had previously applied for three permissions to drill at Shepherdswell, Guston and Tilmanstone in 2013, but these were all withdrawn before being determined. However, an e-mail from the company's Chief Executive, Gerwyn Williams, leaked to the East Kent Against Fracking group revealed that the license blocks for all the holdings in East Kent had been surrendered except for the area around Tilmanstone, and that the company plans to drill an exploration well through the six deepest coal seams. Mr Williams advised that he will be contacting the County, District and Parish Councils about the plans before holding a public meeting.

Better news for the anti-fracking community came on the 9th November when a resolution proposed by Westhoughton Town Council was passed at the Annual General Meeting of the Lancashire Association of Local Councils (LALC). LALC is the professional representative body for 187 (out of 215) Town and Parish Councils in Lancashire, and is also a member of the National Association of Local Councils (NALC). The resolution, which was overwhelmingly approved, stated:

> "That L.A.L.C. and N.A.L.C. urge individual Parish & Town Council's [sic] to oppose applications for fracking in their areas, in recognition that the potential damage to the environment is irreversible and no payment from

fracking companies can compensate for any such damage."

Figs.37 and 38: The drill site at Ellesmere Port, 14th December 2014

One of the Councils to have expressed concern at fracking, Cheshire West and Chester, began legal actions in September 2014 to evict a Community Protection Camp at an IGas exploration site at Ellesmere Port. The Council gave permission for a single, vertical borehole at the site and the campaigners argue that the operation could likely lead to hydraulic fracturing in the near future. Despite having previously expressed concern at fracking, a spokesman for the Council said that "[we have] no official policy on this issue and at present an all party working group on unconventional gas and oil extraction is considering evidence from all sides involved in this debate."

Walk the Walk '14

Gayzer Frackman, who had previously mounted a protest walk from Blackpool to Downing Street in 2012, performed another ambitious journey on foot against the shale industry in November 2014. This time, he was walking in support of a Romanian campaigner, Alexandru Popescu, who was also embarking upon an epic walk. Popescu had undertaken a hunger strike during the winter of 2013 against the expanding shale industry in his home country and decided to embark upon a three-month long, 1,957 mile (3,150km) walk on the 31st August 2014 from Romania to Brussels where he delivered a statement to the European Parliament against fracking in Europe on the 1st December – Romania's national day.

On the 17th November, Gayzer (accompanied by his meerkat mascot, Fracky) set off from Downing Street on his way to meet with Alexandru outside the European Parliament, via France, to give a message calling out against fracking. The pair had already gained the support of several MEPs, including Keith Taylor – the Green Party MEP for the South East of England – in advance of reaching their destination. The event has inspired many people across Europe in the fight against fracking.

Autumn 2014: Three Warnings in One Month

Earlier in the year, anti-fracking campaigners launched a series of five community blockades in a matter of four months against exploratory drilling sites across the UK. Another important wave of activity came in the autumn, when three independent and influential reports were published within a month of each other. The first of these reports was a health impact assessment by the Director of Public Health for Lancashire County Council in the final week of October. This advised that the potential for two fracking sites in the county (Cuadrilla's applications for Preston New Road and Roseacre) was already causing an array of health conditions in local residents, including "loss of control over life, and feeling there is no escape; helplessness, and an inability to change the outcome of the decision-making process; disenfranchisement; inability to protect themselves or take responsibility for their health; anxiety, and facing the unknown; stress; sleep disturbance; depression." Eleven conclusions are made in the report, ranging from involving local residents and making the process much more transparent and inclusive. Although the report was far from suggesting that fracking be banned, it did acknowledge that "although the exploration of shale gas is temporary, it is not short term" and highlights that each of the two proposed Lancashire sites could take up a year each to drill and frack.

The second report was announced by Bianca Jagger on the 31st October, and published on the 26th November. Commissioned by the Bianca Jagger Human Rights Foundation the report, entitled, A Human Rights Assessment of Hydraulic Fracturing and Other Unconventional Gas Development in the United Kingdom, was delivered to No.10 Downing Street and called for an immediate moratorium on hydraulic fracturing until a full and publically-funded human rights impact assessment had been undertaken by the Government. It detailed several risks of unconventional oil and gas, including water contamination, radiation, impact on air quality, seismic events and impact upon climate change. It compared them to the human rights to life and security of person, right to water and health, right to home and private life and the right to public participation in the decision making processes.
Launching the report, Ms Jagger said that "The Government has disregarded the Human Rights of ordinary citizens" in the dash for gas by promoting fracking in spite of "well-documented health and environmental impacts", by changes to the trespass law and through the Infrastructure Bill.

The third and final report came to public attention when published on the 28th November, when the Government's Chief Scientific Advisor's annual report warned that the degree of risks from fracking are as serious as those from thalidomide, tobacco and asbestos in a chapter written by Professor Andrew Stirling of the University of Sussex.
It also argued that the world could tackle climate change through energy efficiency measures and renewable technology, but warned that "high-profile self-fulfilling

assertions to the contrary, including by authoritative policy figures" and vested interests in the success of the fossil fuel industry stood in the way of this. Campaigners have hailed this report as a "naked-emperor moment for the government's dash to frack" and are urging Ministers to take note of the advice from their own Chief Scientific Advisor.

Frack Attack – Round Three

The news broke on the 25th November 2014 that Third Energy (which is 97% owned by Barclays Natural Resources Investments, a private division of Barclays Bank) are to apply for permission to undertake hydraulic fracturing near the village of Kirby Misperton in North Yorkshire. Though the announcement was a surprise, it was somewhat expected by the anti-fracking community. Suspicion was first roused when Viking UK Gas Ltd (a subsidiary of Third Energy) submitted an application to North Yorkshire County Council on the 2nd October 2012 for the extension to an existing well site and the drilling of two production boreholes within a designated Area of Outstanding Natural Beauty around 2km (1.2 miles) south of Kirby Misperton and the famous Flamingo Land Resort. One of the wells was targeting a conventional gas reserve, but examination of the second well revealed that it was to be drilled deeper to around 2,743m (9,000 feet) and descend into the Bowland Shale layers.

Despite this, the planning statement noted that "The drilling is targeting conventional gas bearing formations", though did admit that "The Bowland Shale…may be cored to analyse the reservoir potential of the zone." Permission was subsequently granted on the 9th January 2013.

Further suspicion was raised, when the drilling of the supposed 'conventional exploration' well began in September 2013, when it was noticed that the rig – a Drillmec HH220 – belonged to Cuadrilla and was the very same one used to drill the fracking well at Preese Hall in Lancashire. The logo had been crudely taped over, but closer inspection showed that part of it was still visible. It was later revealed that samples of the shale layers had been taken for assessment.

In response to the news by Third Energy, the local MP, Anne McIntosh, issued a statement:

> "I am at a loss to understand how Third Energy have suddenly developed the technology [to frack], when they gave me an assurance within the last year that they themselves neither had the technology nor any intention of hydraulically fracking at depth anywhere in Ryedale."

On the 16th December, Ms MacIntosh requested that David Cameron fully disclose a report by DEFRA into the impacts of shale gas upon the rural economy that was heavily redacted when released earlier in the year, and challenged the Prime Minister to change rules that allow drilling companies to self-regulate.

Aside from causing huge concern amongst the local residents, the news has also reinforced the concern of anti-fracking campaigners at the other end of the country. A similar situation has been developing at Broadford Bridge, near Billingshurst in West Sussex, where Celtique Energie and their financial partners, Magellan Petroleum, are proposing to core and log the Kimmeridge and Liassic Shale layers whilst drilling a 'conventional' exploration well and saying that there would not be any hydraulic fracturing. Furthermore, amongst the documents submitted by Celtique during the application process, it was stated that the intended drilling rig will be a Drillmec HH220 - the same type as used at the Preese Hall fracking well, and during the initial drilling at the Kirby Misperton Deep Well.

5
OTHER KEY POINTS OF THE FRACK FREE MOVEMENT

Having taking a journey through the anti-fracking movement, from its early and tentative beginnings to the high-profile national force that it has become, there are a number of key themes that have been persistent throughout the story, or otherwise have great significance to the movement either through support of the campaigning or other threats related to fracking. In this chapter we will explore a number of other key points that have been important in the story of the UK Anti-Fracking Movement.

Protest Policing and Covert Intelligence
Ever since the Balcombe Blockade, the behaviour of the various police forces have been called into question innumerable times by the anti-fracking community, social welfare groups and libertarian organisations alike, usually in response to arrests and covert operations. The use of so-called 'snatch-squads' was a frequent sight at the Balcombe Blockade in 2013, and were the subject of much criticism. Snatch-squads are small groups of four or five police officers singling out a person for immediate arrest without warning, quickly removing them to an awaiting van and driving off to a Police Station. Defending this tactic, Sussex Police have claimed that far from being random arrests of protestors, the 'snatch squads' are used to make an arrest on a person suspected of having committed an offence hours, or even days, beforehand, but could not be apprehended at the time. However, criticism surrounds claims of heavy handedness, specific targeting of those seen to be leaders and random arrests designed to discourage participation in any protests. The accusation of discouraging participation – or criminalising peaceful protest–

has been a mainstay in opinion throughout the anti-fracking movement during 2013 and 2014. Speaking out against arrests made at Balcombe, the barrister representing many of those charged with offences said:

> "What they [the police] did criminalised protest. They used the section 14 orders and bail conditions, which were imposed on everyone, and which stopped them from going within miles of the site, to stop them from protesting. It was like an injunction by the back door. If you turn up – new to protest – and...then see people being arrested and handcuffed, it is quite shocking and frightening and puts you off being there."

The large police presence, said to be highly disproportionate, is also a common cause of complaint at the various protest sites around the country, as are the high costs associated. At Balcombe, Sussex Police faced a bill of nearly £4 million and many called for Cuadrilla to bear the costs since it was their presence and activity that was responsible for the protest occurring. Similar calls were made for the policing operation at Barton Moss. Early arrests at Balcombe often had "draconian bail conditions" placed upon suspects that prevented them from entering a large defined zone around Balcombe covering tens of square miles of West Sussex and even stopped them travelling on the Brighton to London mainline train service. Although many of these over-the-top conditions were dropped when challenged, on the grounds of limiting the right to peaceful protest, the condemnation of tactics also included the use of mass arrests, unlawful arrests and misuse of section 14 notices under the Public Order Act 1986. These notices are used to essentially prohibit a designated area to anyone wishing to protest, whether it be attempting to block access to a site, or simply holding a placard at the side of the road. These were dismissed by Brighton Magistrates Court rulings, that highlighted the fact that the Public Order Act specifically includes the word 'serious' in its content with regards to senior officers having to prove that in their belief a protest would risk serious disorder, damage or disruption to the wider community for a section 14 notice to be valid. An investigation into the issue of pre-charge bail by The Guardian in late December 2014 uncovered the fact that 85.6% of all those prevented from attending protests since 2008 were never charged with having committed any crime, and suggested that the tactic could be used as a means of 'muzzling' people and disrupting protest activity. The Metropolitan Police have issued the highest number of bans with 569 people prohibited from protesting, followed by Nottinghamshire Police (120), City of London Police (45), Essex Police (39) and Sussex Police (36).

Of the 126 people arrested at Balcombe, almost a third had the charges dropped by the police service before reaching court and only a quarter of those taken to court were found guilty of an offence – 75% of all the arrests were wrongfully made or otherwise unlawful. In addition to this, accusations were made against individual members of Sussex Police for a variety of reasons. One incident involved an Officer using abusive language against the protestors at Balcombe, branding them as "scum" on Twitter. The Officer was identified and subsequently received "management advice" about their behaviour from Sussex Police. Other incidents include a number of Officers removing or otherwise obscuring their identification numbers whilst making arrests, and even the wilful breach of regulations and/or the law on more than one occasion. For example, early on in the protest, a film emerged on YouTube that apparently showed members of G4S – the private security firm used by Cuadrilla – making contact with protestors who had locked-on to the entrance gates. As G4S staff attempted to remove their restraints, a number of Officers watched on, despite being informed that it was against the law for the security personnel to make physical contact with people if they were not posing a threat to people or property. Several times the police were informed of alleged breeches of the law by site workers, such as using mobile phones whilst driving and even the attempted ramming of one person by a site worker driving a car, but are accused of taking no action.

By far the largest target of complaints and criticism was Greater Manchester Police – in particular their Tactical Aid Unit – at the Barton Moss protest in late 2013 and early 2014. There were a number of reported incidences of people having been assaulted or otherwise injured, allegedly at the hands of the police during arrests. During one lorry escort, a video appears to show an elderly and disabled man being pushed into a steep sided ditch as Officers rushed forward to arrest a pregnant woman. This resulted in him requiring medical attention. Another time, a veteran of the Balcombe Blockade suffered a fractured eye socket and other facial injuries allegedly received during an extremely violent arrest.

A number of times, photos and films have appeared that seem to show arrested protestors with blood clearly visible on their faces, such as one woman who required hospital treatment after being allegedly strangled and dragged by her handcuffs during an arrest, photographers being assaulted and another man requiring hospitalisation after suffering a broken leg.
In one shocking case, a 15-year-old girl was allegedly singled out and arrested on her first visit to the camp during research for a school geography project. She was separated from her family and taken to the Police Station without an accompanying adult. Greater Manchester Police were also accused of fabricating evidence to support their arrests. In the first three months of the camp, almost all arrests were on the charge of obstructing the highway, but in a civil case against these charges in February 2014 a photograph emerged apparently showing police officers in early January removing

signs indicating that the 'road' was in fact a footpath. On the 13th February the District Judge ruled that the 'road' in question was in fact a public footpath and not a highway, making some 50 arrests unlawful. On one day of the protest, Police closed the footpath and barred any access by legal observers or members of the press to those trapped inside the cordon. It is alleged that those press and legal observers inside the cordon were arrested when attempting to film the actions of the police. On the 3rd July 2014, a Deputy District Judge at Manchester Magistrates Court on discussing the relationship between IGas and Greater Manchester Police (GMP) amongst accusations of collusion between the two bodies stated that the police had on several occasions exceeded its powers by intervening on IGas' behalf during civil trespass and accused GMP of "acting as civil enforcement officers" for the company.

Fig.39: Police Evidence Gatherers have become a common sight at political protests in the UK

From the very beginning of the anti-fracking movement it has been an expectation that undercover police, the intelligence services and even agent provocateurs would infiltrate the campaign. This is why the movement has been organised in such a way so as that there are no 'leaders' or persons in positions of power or control, in order to limit as far as possible the opportunities for persons unsympathetic to the movement from gaining undue influence that could damage the campaign. Whilst this might appear far fetched at first sight, there have been a number of UK campaign groups in the recent past that have been infiltrated in this manner, with legal proceedings exposing undercover agents as playing a key part in wilfully encouraging the organisations and their members to undertaking unlawful actions. In some cases the agents themselves were charged with exceeding their powers and duties and inciting illegal behaviour designed to discredit other people or campaigns. Furthermore, in June 2014 a report written by Hertfordshire and Essex Police on the Balcombe policing operation released under a Freedom of Information Act request revealed that "covert tactics" including the use of undercover Officers amongst the protestors were used in 2013. The sections referring to these secretive operations were redacted, but in a botch by the police the redacted sections could be uncovered by simply highlighting the relevant text, or by copying and pasting the document into a word processing programme. Responding to these revelations, Caroline Lucas MP said "I am deeply concerned about the waste of police resources going on here [at Balcombe] and how legitimate peaceful protest is being criminalised in this way".

Fig.40: Police presence at anti-fracking protest has become bolstered by high numbers of intelligence units

The matter of intelligence-gathering extends beyond just protests. In early February 2014 it was revealed in a meeting agenda for a Pulborough Parish Council meeting that the Parish Office, and the offices of those Councils in the surrounding parishes of West Chiltington and Billingshurst in West Sussex, had been visited by Sussex Police in a request "to advise them [the Police] of contact from residents" regarding the nearby Broadford Bridge drilling site that is located near the intersection of the boundaries of all three parishes.

Although having made no formal representations on the Broadford Bridge site, the author was amongst those whose names were mentioned to the police during the visit. Furthermore, it is alleged that on the days leading up to the protest march in late 2013 at nearby Wisborough Green and Kirdford, that the police also made visits to local businesses to advise them to expect possible trouble and disruption (even though many of those visited took part in the march). This was added to in early 2014 when e-mails were sent to businesses and other organisations in Pulborough and Wisborough Green from Sussex Police requesting any information they might have on what impact any protest activities at Celtique Energie's Broadford Bridge and Wisborough Green sites would have upon traffic, businesses and residents in their areas.

Furthermore, Canterbury Christ Church University decided to host a public debate on fracking on the 19th November 2014 which was attended by around 200 members of the public who had to book the tickets in advance. However, on the 15th December The Guardian broke the news that Kent Police had asked the University shortly before the meeting to hand over a list of those planning on attending the debate, though the University refused to give the information. Widespread outrage was understandably the reaction to the news, though the situation was aggravated when the Kent Police and Crime Commissioner stated in an article on the 17th December that upon questioning the Chief Constable that no such request had been made to the university. This was despite the fact that Kent Police had previously admitted that the list of names was asked for and that the University had also confirmed that a list of names of attendees was requested. When the information was not supplied, an Inspector was sent to attend the meeting "as an interested stakeholder". The newspaper also ran an article on the 2nd December reporting that the President of Lancaster University's Students' Union had posters displayed in her office window – including a 'Not for Shale' sign – photographed by police officers attending the University for Freshers' Week. When questioned, the officers replied that a Public Order offence was potentially being committed by displaying the posters.

In response to the policing of anti-fracking protests since Summer 2013, the Network for Police Monitoring ('Netpol') was awarded a grant from the renowned Joseph Rowntree Reform Trust in order to monitor and assess the policing of anti-fracking protests over the next two years from October 2014. Netpol state that the funding:

> "will enable us to use the evidence we gather to campaign for a less antagonistic policing strategy towards opponents of unconventional energy extraction…Netpol intends to actively support for the right of local people and their allies to protest their opposition to fracking. There is no reason why opponents should inevitably expect to face disproportionate oppressive policing and an increased risk of criminalisation just because their target is an expanding and lucrative industry with considerable government support."

The organisation aims to:- assess the developing tactics used by the police at protests; the numbers and types of arrests and charges made and common complaints of police violence or misuse of powers; identify the extent of police surveillance and any tactics that impact upon the right to protest; and to brief solicitors and defence barristers of emerging police strategies and their political implications. More information on policing of protests can be found on the Network for Police Monitoring website.

NATO Enters the Debate

Contrary to the assertions that the UK anti-fracking movement is a target for undercover operatives seeking to manipulate the campaign, the outgoing Secretary-General of NATO (the Euro-American military organisation), Anders Fogh Rasmussen, accused the anti-fracking movement of either being infiltrated or working in partnership with Russian secret agents. Speaking in June 2014, he said:

> "I have met allies who can report that Russia, as part of their sophisticated information and disinformation operations, engaged actively with so-called non-governmental organisations – environmental organisations working against shale gas – to maintain European dependence on imported Russian gas…That is my interpretation".

Supporting the allegation, the Chief Executive of the UK Onshore Operators Group, which is calling for the roll out of full scale hydraulic fracturing and other unconventional oil and gas technologies, said "If it is true that Russia is funding anti-fracking groups, it deserves a full investigation and disclosure of their sources of support". Responding to this and Mr Rasmussen's allegations, the head of Friends of the Earth stated "Perhaps the Russians are worried about our huge wind and solar potential, and have infiltrated the UK government." He was perhaps making reference to the apparent increase in Government opposition to renewable technologies, including removing its support of onshore wind turbines and reducing financial support into renewables research whilst providing tax breaks to fracking companies, believed to be in response to the rise of UKIP as a mainstream political party. UKIP, the largest threat to Conservative votes, openly deny the notion of man-made climate change, oppose wind and solar renewable energy technology and investment and are avid supporters of shale gas. In September 2013 the party's energy spokesperson, Roger Helmer MEP, stated in his speech to the UKIP annual Conference "I have absolutely no sympathy for the rent-a-mob protestors, the Swampies and the Occupy Movement and the anti-capitalists and eco-freaks who have sought to hijack the Balcombe protest." He went on to chastise famous names such as Dame Vivienne Westwood and Bianca Jagger, as well as the RSPB, for their anti-fracking stance. Similar unfounded accusations to those made by the NATO Secretary-General against the movement appeared very early on when the Lancashire-based Ribble Estuary Against Fracking group were accused in 2011 by some pro-shale lobbyists of being infiltrated by anarchists.

Whilst the industry are inclined to be supportive of the somewhat absurd notion that thousands of ordinary men, women and children across the UK are in collusion with the Putin Regime, calls by frack free campaigners for investigations into funding and influence by the oil and gas industry upon Government and Council policies and voting behaviour have gone unanswered.

Council Investments

Widespread condemnation has come from the anti-fracking movement since late 2013 over revelations that many of the Mineral Planning Authorities in England who are dealing with applications for oil and gas drilling are also invested in those very same companies seeking permission in their areas. Beginning in October 2013, the Kent branch of the Green Party of England and Wales revealed through a Freedom of Information Act request that the County Council has £153 million worth of shares in companies involved in extracting shale oil and gas. This was followed by members of campaigning group Keep Kirdford & Wisborough Green in early 2014 revealing via the same method that the West Sussex County Council Pension Fund holds more than £82.5 million directly invested in fossil fuel companies, including over £66.3

million in those partly or wholly engaged in fracking, and an additional £3.8 million held in Cuadrilla's UK fracking investment partner Centrica. The Pension Fund also holds several tens of millions of pounds of investments in the arms and tobacco trades. Building on this, The Ecologist Magazine also revealed that West Sussex County Council's Pension Fund also has investments in Cuadrilla, IGas and Celtique Energie – all of whom either operate or are applying to operate sites in West Sussex. The IGas investments are via an undisclosed percentage of a massive £187 million stake in fund managers, Ballie Gifford. The Ecologist also revealed similar investments in these companies by the Greater Manchester Pension Fund, where IGas had received permission to drill a combined CBM and shale gas exploration well.

These investments have led to calls by campaigners to remove the power of the councils to determine oil and gas applications and instead give local communities the chance to decide whether to allow drilling in their areas via referenda, citing a conflict of interest existing due to the sums involved. Whilst the councils in question deny that the investments impact in any way upon the planning process, criticism remains that they undeniably stand to benefit from the success of shale gas in the UK and also have the legal duty to see the best financial return on their investments.

European Union and Shale Gas
In the previous European Parliamentary session prior to the 2014 European Elections (2009 to 2014), there were eight votes on matters regarding hydraulic fracturing. Below are the voting records of the 15 UK political parties that were represented in the previous Parliament on each of the eight motions. A brief summary of what each motion called for is also provided.

1. A Europe-wide Moratorium on the use of Hydraulic Fracturing

Because of the adverse climate, environmental and health impacts and the gaps in the EU regulation, hydraulic fracturing should not be authorised by EU Member States.

	For	Against	Abstain	Absent
An Independence Party		100%		
British Democratic Party		100%		
British National Party (BNP)	100%			
Conservative Party		92%		8%
Democratic Unionist Party (DUP)		100%		
Green Party	100%			
Independent/No Party Affiliation		50%		50%
Labour Party	62%	15%	23%	
Liberal Democrats	33%	67%		
Plaid Cymru	100%			
Scottish National Party (SNP)		100%		
Sinn Féin	100%			
United Kingdom Independence Party		89%		11%
Ulster Conservatives and Unionists		100%		
We Demand A Referendum Party		100%		

2. Strict Rules for Hydraulic Fracturing in Europe

The report recommends a series of measures that would set very high standards for any hydraulic fracturing operation in Europe.

	For	Against	Abstain	Absent
An Independence Party		100%		
British Democratic Party		100%		
British National Party (BNP)	100%			
Conservative Party		92%		8%
Democratic Unionist Party (DUP)		100%		
Green Party	100%			
Independent/No Party Affiliation			50%	50%
Labour Party	69%		31%	
Liberal Democrats	100%			
Plaid Cymru	100%			
Scottish National Party (SNP)			100%	
Sinn Féin		100%		
United Kingdom Independence Party		11%	78%	11%
Ulster Conservatives and Unionists		100%		
We Demand A Referendum Party		100%		

3. Mandatory Environmental Impact Assessments for Shale Gas Exploitation

Environmental Impact Assessments (EIAs) are required for the exploitation of shale gas and other unconventional fossil fuels in Europe.

	For	Against	Abstain	Absent
An Independence Party		100%		
British Democratic Party		100%		
British National Party (BNP)				100%
Conservative Party		80%		20%
Democratic Unionist Party (DUP)		100%		
Green Party	100%			
Independent/No Party Affiliation		50%		50%
Labour Party	92%			8%
Liberal Democrats	8%	75%		17%
Plaid Cymru	100%			
Scottish National Party (SNP)	100%			
Sinn Féin	100%			
United Kingdom Independence Party		67%	11%	22%
Ulster Conservatives and Unionists		100%		
We Demand A Referendum Party			100%	

4. Mandatory Environmental Impact Assessments for Shale Gas Exploration

EIAs are required for the exploration and evaluation of shale gas and other unconventional fossil fuels in Europe.

	For	Against	Abstain	Absent
An Independence Party		100%		
British Democratic Party				100%
British National Party (BNP)				100%
Conservative Party		80%		20%
Democratic Unionist Party (DUP)		100%		
Green Party	100%			
Independent/No Party Affiliation		50%		50%
Labour Party	85%			15%
Liberal Democrats		83%		17%
Plaid Cymru	100%			
Scottish National Party (SNP)	100%			
Sinn Féin	100%			
United Kingdom Independence Party		67%	11%	22%
Ulster Conservatives and Unionists		100%		
We Demand A Referendum Party			100%	

5. Mandatory Environmental Impact Assessments For Shale Gas Exploitation and For Exploration When It Involves Hydraulic Fracturing

EIAs are required for the exploitation of shale gas and other unconventional fossil fuels in Europe and for the exploration if it involves hydraulic fracturing.

	For	Against	Abstain	Absent
An Independence Party		100%		
British Democratic Party	100%			
British National Party (BNP)				100%
Conservative Party		80%		20%
Democratic Unionist Party (DUP)		100%		
Green Party	100%			
Independent/No Party Affiliation		50%		50%
Labour Party	92%			8%
Liberal Democrats	75%		8%	17%
Plaid Cymru	100%			
Scottish National Party (SNP)	100%			
Sinn Féin	100%			
United Kingdom Independence Party		67%	11%	22%
Ulster Conservatives and Unionists		100%		
We Demand A Referendum Party			100%	

6. Call on Local Authorities to Ban the Exploitation of Unconventional Fossil Fuels

Because the EU cannot decide on whether or not to ban unconventional fossil fuel exploitation or hydraulic fracturing, local and regional authorities are strongly encouraged to do so.

	For	Against	Abstain	Absent
An Independence Party		100%		
British Democratic Party		100%		
British National Party (BNP)	100%			
Conservative Party		92%		8%
Democratic Unionist Party (DUP)		100%		
Green Party	100%			
Independent/No Party Affiliation		50%		50%
Labour Party	54%	38%		8%
Liberal Democrats	25%	58%		17%
Plaid Cymru	100%			
Scottish National Party (SNP)				100%
Sinn Féin	100%			
United Kingdom Independence Party		67%	11%	22%
Ulster Conservatives and Unionists		100%		
We Demand A Referendum Party			100%	

7. Call on the European Commission to Include Mandatory Environmental Impact Assessments For Shale Gas Projects

Calls on the European Commission to include mandatory EIAs for both exploitation and exploration in the legislative work on shale gas.

	For	Against	Abstain	Absent
An Independence Party				100%
British Democratic Party		100%		
British National Party (BNP)		100%		
Conservative Party		84%		16%
Democratic Unionist Party (DUP)		100%		
Green Party	100%			
Independent/No Party Affiliation		50%		50%
Labour Party	100%			
Liberal Democrats	92%			8%
Plaid Cymru	100%			
Scottish National Party (SNP)				100%
Sinn Féin	100%			
United Kingdom Independence Party		78%		22%
Ulster Conservatives and Unionists		100%		
We Demand A Referendum Party		100%		

8. Transparency For the Chemicals Used in Hydraulic Fracturing

Calls on the European Commission to pass legislation for ensuring the highest possible level of transparency regarding the chemicals used in hydraulic fracturing.

	For	Against	Abstain	Absent
An Independence Party				100%
British Democratic Party	100%			
British National Party (BNP)	100%			
Conservative Party	4%	76%		20%
Democratic Unionist Party (DUP)		100%		
Green Party	100%			
Independent/No Party Affiliation		50%		50%
Labour Party	100%			
Liberal Democrats	59%		33%	8%
Plaid Cymru	100%			
Scottish National Party (SNP)				100%
Sinn Féin	100%			
United Kingdom Independence Party		78%		22%
Ulster Conservatives and Unionists				100%
We Demand A Referendum Party		100%		

The voting records of the UK MEPs tend to reflect how each party stands with regards to shale gas and hydraulic fracturing back in the UK. Of the main parties, only the Green Party are wholly opposed to fracking whereas both UKIP and the Conservative Party are keen for hydraulic fracturing to be rolled out nation wide as quickly as possible with the least amount of regulation on the industry. Both Labour and the Liberal Democrats are in favour of hydraulic fracturing, but also support stringent regulation of the industry to help protect health and the environment. The Liberal Democrats, however, are more inclined to favour a faster roll out at the expense of strict governance. Of the regional parties, the Welsh Plaid Cymru Party is completely opposed to fracking, as is the Irish Sinn Féin. The other two Northern Irish parties represented in the European Parliament are extremely keen to see fracking take place. In Scotland, the SNP are not entirely opposed to fracking, but recognise that if it takes place then there would need to be tight regulatory control of the industry.

One of the main arguments used by the anti-fracking movement is the lack of information about the chemicals being used in drilling operations. For example, although pro-shale advocates claimed Cuadrilla to be using only non-hazardous, household additives in their drilling fluid at Balcombe in 2013, later Freedom of Information Act requests to the Environment Agency revealed that the company had in fact been given permission to use Ethylene Oxide – also known as Oxirane – and had applied for permission to use Antimony Trioxide. Both are certainly not household chemicals and are considered to be toxic substances. Additionally, a later Freedom of Information request revealed that more than 30 different chemicals have been permitted for use at the nearby Broadford Bridge site of which, according to the **Material Safety Data Sheets**, six are known or suspected carcinogens and more than 74 health warnings are attached.

However, of most concern based upon the EU voting records is that four of the fifteen political parties, including UKIP and the Conservatives, are opposed to publicising the chemicals to be used in hydraulic fracturing. Additionally, the Liberal Democrats only voted in favour of this particular motion by a narrow margin – one third of all their MEPs voted to abstain. Notable politicians who voted against the motion include the UKIP leader, Nigel Farage, and the UKIP energy spokesperson, Roger Helmer. The pair also voted against measures to introduce "very high standards" for fracking operations whilst Roger Helmer voted to oppose all the votes calling for mandatory EIAs for shale gas.

The Transatlantic Trade and Investment Partnership (TTIP)

Currently under negotiation at the time of writing is an international free trade agreement between the EU and USA, known as the Transatlantic Trade and Investment Partnership, which is taking place largely behind closed doors. It covers a wide range of subjects such as food, genetically modified organisms, energy and data protection. Of major concern to the anti-fracking movement (and many other non-governmental organisations) is that the TTIP includes a clause that in effect would allow American companies and/or their investors to sue European governments and have policies overturned if it is deemed to have a reducing effect upon profitability. Situations such as this are already occurring with the EU-Canada Comprehensive Economic and Trade Agreement (CETA). The fear regarding fracking is that American companies would be entitled to legally challenge and overturn any European bans on the technology, such as are in place in France, Germany and Spain, or even remove and amend industry regulations and laws if they are seen as hindering the profitability for investors. As with the wider subject of hydraulic fracturing, only the Green Party are opposing the TTIP whereas Labour, Liberal Democrats, UKIP and the Conservatives are supportive of the negotiations.

The Infrastructure Bill 2014

Of more immediate concern to the potential for a UK-wide roll out of hydraulic fracturing is the Government's Infrastructure Bill. Amongst the new regulations that the Bill proposes to implement are the removal of landowner and householder rights for underground access and allow drilling companies to operate below 900 feet (300m) without the need for permission as well as measures to speed up the planning process for "nationally significant projects" – this includes hydraulic fracturing. Whilst these proposals have been known about since the Queen's Speech in 2014, what is of the greatest concern to frack-free activists is that a report on the Bill contains an entire section devoted to fracking which has been left completely blank, despite the Bill being in the advance stages of passage through the House of Commons. Many have raised the question about how can a nationally significant Bill such as this be voted upon and potentially enshrined in law when entire sections are left blank, and not to be written until after the Bill has been passed – the MPs and Lords are essentially debating and voting upon the piece of legislation with no knowledge of its contents! On the 26th September, the Department for Energy and Climate Change announced the results of the consultation on underground access – some 40,647 responses were made with an astounding 99% rate of objection to the plans to allow drilling without landowner permission. However, in a press release DECC stated that:

> "We acknowledge the large number of responses against the proposal and the fact that the proposal has provided an opportunity for the public to voice their concerns and raise issues. However the role of the consultation was to seek arguments and evidence to consider in developing the proposed policy. Whilst a

wide range of arguments were raised and points covered, we did not identify any issues that persuaded us to change the basic form of the proposals."

Widespread condemnation of this disregard for the consultation results was immediate. Further condemnation came on the weekend of the 11th – 12th October 2014 when Baroness Kramer introduced a post-consultation amendment to the Bill just one working day before the scheduled House of Lords debate which would permit companies to "[pass] any substance through, or put any substance into, deep-level land, or infrastructure installed in deep-level land" and "leave deep-level land in a different condition from the condition it was in before an exercise of the right of use (including by leaving any infrastructure or substance in the land". Both pro- and anti-fracking landowners and homeowners have expressed deep concern at this latest attempt to remove ancient Common Law rights. They claim that it would leave companies free from responsibility for any future contamination issues, and would also allow companies to inject radioactive material from fracking operations and nuclear energy production under homes and land without permission.

Having now spent some time examining other key issues that go against or otherwise hinder the aims and objectives of the anti-fracking movement, these final few points take a look at some of the significant factors that are helping to raise awareness and support the fight against fracking in the UK.

Druids Against Fracking and The Warriors Call
Like the Knitting Nanas, a common and welcome sight at many of the community protection blockades and protest marches across the UK have been a group of people who have come together under the banner of 'Druids Against Fracking'

and 'The Warriors Call'.

Both organisations are a collective of people who have been called to follow the ancient pagan belief systems that honour the earth and value all living beings and the nature spirits. The pagans and druids believe that hydraulic fracturing is a "violent assault upon Mother Nature and all her inhabitants", as well as sharing all the many other concerns held by other religious and non-religious anti-fracking activists. Often conducting meditations and protection rituals, many find the groups a calming and reassuring presence at demonstrations and community blockades. Shortly after the Balcombe Blockade of 2013, a special anti-fracking protection ritual was held at Glastonbury Tor on the 28th September which was attended by over 1,000 participants from pagan and 'new age' belief systems, that called for protection against the threat of fracking in the UK. At the same time, hundreds of other similar rituals were taking place across the world, each asking that their homelands be safe from the drills. What many might find astonishing is that this world-wide union was not an organised event, but occurred spontaneously as a collective response to threats to Mother Earth.

Wrong Move Campaign
We have already briefly visited Greenpeace's Wrong Move campaign, but it is worth a more detailed look at this innovative and significant movement that has already had deep repercussions for planned drilling operations at Fernhurst in West Sussex. Whilst its name is a parody of the Right Move property selling website, the Wrong Move Campaign aims to help raise awareness of those areas of the UK at risk from potential fracking operations, and assist those opposed to the technology to challenge companies with licences to drill in or near their communities. The Wrong Move name actually refers specifically to the online hub for the Not for Shale legal blockade that allows people to

expressly deny permission to drill under their land or homes, even if operations are planned or already taking place. So far, over 49,000 people have signed up to the legal block and at Fernhurst, the proposed Celtique drill site has been completely surrounded by blockaders to the extent that a planned horizontal well has had to be dropped from the plans and questions remain regarding access to the site.

Talk Fracking

As a side branch to the anti-fracking movement, a related campaign has been underway since June 2014 to get the nation debating the pros and cons of hydraulic fracturing. The campaign began with a collective letter signed by 150 celebrities, academics, charities and businesses that was sent to the Prime Minister urging an immediate and complete ban on fracking until a full, balanced national debate has taken place. The signatories were:

Academics, Scientists, Lawyers and Activists

Dr Laura Adams, Professor Erik Bichard, John Christensen, Professor Lawrence Dunne, James Hansen, Mike Hill, Bianca Jagger, Bruce Kent, Dr David P Knight, Professor Sir Harold Kroto FRS, Dr David Lowry, Dr Caroline Lucas MP, Dr George Manos, Michael Mansfield QC, Bob Marshall-Andrews QC, Dr Hugh Montgomery, Richard Murphy, Baron Rea of Eskdale, Dr Damien Short, Professor David Smythe, Peter Tatchell, Chris Venables and Professor Graham Warren.

Organisations

Bill McKibben (350.org), End Ecocide EU, Environmental Justice Foundation, Friends of the Earth, Fuel Poverty Action, Gaia Foundation, Greenpeace, Royal Society for the Protection of Birds (RSPB), Salmon and Trout Association, Tracey Marchioness of Worcester and Young Friends of the Earth.

Trade Unions
Chris Baugh (Assistant General Secretary PCS), Manuel Cortes (General Secretary TSSA) and Stephen Hedley (Assistant General Secretary RMT).

Chefs, Food and Farming
Gabriele Corcos, Fergus Henderson, Mark Hix, Hop Fuzz Brewery, River Cottage, Sam & Sam Clark, Geetie Singh and Guy Watson.

Authors, Journalists and Filmmakers
Guillem Balague, Alistair Beaton, Rosie Boycott, Philip Carr-Gomm, Daryll Cunningham, Mark Ellingham, Mariella Frostrup, Neil Gaiman, Jonas Grimas, Mark Haddon, Naomi Klein, Avi Lewis, George Monbiot, Alan Moore, New Internationalist, Dana Nuccitelli, Deborah Orr, John Pilger, Will Self, Nicholas Shaxson, Chris Stewart and Jeanette Winterson OBE.

Fashion
Lily Cole, Bella Freud, Katherine Hamnett CBE, Georgia May Jagger, Andreas Kronthaler, Stella McCartney, Dr Noki, Alexandra Shulman and Vivienne Westwood OBE.

Theatre, Film and Television
Josh Appignanesi, Bill Bailey, Steven Berkoff, James Bolam MBE, Helena Bonham-Carter, Frankie Boyle, Russell Brand, Sam Branson, Michael Elwyn, Noel Fielding, Stephen Frears, Sadie Frost, Liza Goddard, Lee Hall, Jeremy Hardy, Jonny Harris, Susan Jameson, Baroness Beeban Kidron, Dr Pauline Kiernan, Jude Law, Ken Loach, Matt Lucas, Debi Mazar, Geoffrey Munn OBE, Jenny Platt, Tim Preece, Miranda Richardson, Greta Scacchi, Tracey Seaward, Mark Tildsley, David Yates and Yvonne Walcott Yates.

Photography
Bryan Adams, Willie Christie, Dr Andy Gotts MBE, Mary McCartney, Platon, Juergen Teller, Oliviero Toscani and Andy Willsher.

Music and Radio
Asian Dub Foundation, Carl Barat, Jeff Barrett, Danielle de Niese, Robert del Naja, Paloma Faith, Isabella de Sabata Gardiner, Sir John Eliot Gardiner CBE, Bobby Gillespie, Nick Grimshaw, Debbie Hyde, Chrissie Hynde, Andrew Innes, Geoff Jukes, Sir Paul McCartney, Yoko Ono, Adrian Sherwood, St Etienne and Thom Yorke.

Art
Heather Ackroyd, Jimmy Cauty, Sadie Coles, Tracey Emin CBE RA, Sir Anthony Gormley OBE, Dan Harvey, Mona Hatoum, Michael Landy RA, Saskia Oldewolbers, Cornelia Parker OBE, Jamie Reid, Anne Rothstein, Bob Smith, Roberta Smith, Mark Wallinger, Gillian Wearing OBE RA and Rachel Whiteread CBE.

Business
Vince Adams, Joe Corre, Ben Hopkins, Jeremy Leggett, Lush Cosmetics, Sir Tim Smit KBE, Trillion Fund, Triodos Bank and Dale Vince OBE.

Sport
Dietman Hamann

Beginning on the 9th June 2014, the Talk Fracking road show began the first leg of a nation-wide tour of the UK bringing together expert panellists from both sides of the argument in front of large audiences to debate the pros and cons of hydraulic fracturing and allow the listeners and questioners to make up their own minds about the process. The first event was in Glasgow in the 9th June followed by Nottingham on the

10th, Manchester on the 11th, a tour of Swansea and then London on the 16th June. Whilst representatives of various Government Departments and fracking companies were invited to join the pro panel, only one person responded to the invite. Further tours are expected to be arranged in the future and updates can be found on the Talk Fracking website.

Observer Ethical Awards 2014
Each year the Observer series of newspapers hosts an award show to highlight and celebrate activists engaged in promoting and protecting the environment on both global and local scales. At this year's awards, anti-fracking activists were nominated and shortlisted for two of the three reader-voted categories, including Vanessa Vine (founder of Frack Free Sussex and Britain and Ireland Frack Free!), Anne Power (a prolific anti-fracking campaigner in the North-East) and the Barton Moss Community Protection Camp. The winners were announced on the 13th June 2014 and to the delight of many, Anne Power had been chosen as well-deserved winner of the Local Hero Award.

Ecotricity
Ecotricity is an energy supplier founded in 1996 by Dale Vince OBE (a supporter of the Talk Fracking initiative), becoming the world's first Green Electricity company.
With conventional electricity production being responsible for 30% of Britain's total carbon emissions the aim was to provide an alternative to help reduce this figure. In 2010, the company made the revolutionary step of producing Green Gas – a no-carbon, eco-friendly way of producing gas that can be used to power gas ovens and heating in the home. In addition to being an energy supplier, Ecotricity have developed Britain's first electric super car – Nemesis – which is also a demonstration platform for wind powered vehicles. To support the electric vehicles, the company then launched the Electric Highway

which became the world's first national network of electric vehicle charging stations. With food production being a major contributor of carbon emissions, Ecotricity are also developing concepts such as wind-powered tractors and Farm Energy so that the three largest polluters – energy, transport and food - can be transformed into sustainable, eco-friendly industries. The company works on the model whereby the business pursues outcomes other than simply profit and use the money from customer energy bills to build new sources of Green Energy. In fact, since they have no shareholders to cater for, Ecotricity claims to spend more money per customer on developing new sources of renewable energy than any other British energy supplier. Furthermore, the company offers just one simple tariff for customers, ensuring that it always remains cheaper than the Big-6 energy companies, with no hidden fees or costs. Regarding the anti-fracking movement, Ecotricity have vowed to never use gas sourced from fracking and have offered financial support to all the UK's anti-fracking groups, donating money to the campaign every time a new customer signs up via a link on the various websites. Not only that, but they have also offered to cover the £10,000 court costs for a judicial review against the decision in May 2014 by West Sussex County Council to approve the latest application by Cuadrilla to return to Balcombe for flow testing and flaring.

With our journey through the UK's Anti-Fracking Movement complete, it is now time for you, the reader, to decide whether you support or oppose the exploration and exploitation of unconventional fossil fuels.

APPENDIX ONE

UK ANTI-FRACKING GROUPS DECEMBER 2014

Below is a list of all UK-based anti-fracking groups that exist as of December 2014. The list has been divided into regions for ease of reference.

National and Regional
- Anti-Fracking Network
- Artists Against Fracking
- BIFF! Britain and Ireland Frack Free
- Breathe Clean Air Group
- British Anti Fracking Network
- Businesses Against Fracking
- Community Protection Camps, Travel and Liftshare
- Conservatives Against Fracking
- Druids Against Fracking
- Expats Support Anti-Fracking in the UK
- Frack Free Families (formerly Mothers Against Fracking)
- Frack Free Food Alliance
- Frack Free North West
- Frack Free South East
- Frack Freedom Through Non-Violence
- Frack Freepost Letter Project
- Frack Off (UK)
- FRACK OFF! North West Against Fracking
- Fracking Hell (UK)
- Fracking Nightmare
- Gardeners Against Fracking Forever
- No Fracking in Scotland and the North of England
- No Fracking UK
- No Nuclear Dumping
- Not For Shale
- North East Extreme Energy Awareness
- Protectors' Travelling Fund
- Say No to Fracking Shrewsbury, Oswestry, Wrexham and Chester

- Students Against Fracking
- Sussex Hampshire Awareness Fracking Trust
- The Entire North of England Unite Against Fracking
- The Northern Anti-Fracking Hub
- The Warrior's Call - Pagans United Against Fracking

England
Berkshire:
- Berkshire Against Fracking

Cheshire:
- Anti Fracking Warrington
- Cheshire and Wirral Action on Fracking
- Congleton and Macclesfield Against Fracking
- Ellesmere Port Frack Free
- Elton, Ince, Helsby and Frodsham Local Fracking Info!
- Frack Free Dee
- Frack Free Deeside
- Frack Free East Cheshire
- Frack Free Ellesmere Port
- Frack Free Farndon
- Frack Free Handbridge
- Frack Free Ince, Elton, Helsby & Frodsham & Thornton Le Moors
- Frack Free Malpas
- Frack Free Mickle Trafford
- Frack Free South Wirral
- Frack Free Upton
- I'm Against Fracking at Ince Marshes
- Warrington Against Fracking
- Wirral & Cheshire Awareness Group on UCG, Fracking & CBM
- Wirral Against Fracking
- Wirral Residents' Underground Coal Gasification Awareness

Cleveland:
- Frack Free Cleveland

County Durham:
- Frack Free Stockton

Cumbria:
- Radiation Free Lakeland

Derbyshire:
- Frack Free Derbyshire
- Frack Free Glossop

- Glossop & The High Peak District Against Fracking

Devon:
- Frack Free Devon

Dorset:
- Frack Free Dorset
- Protect our Purbecks
- Protect Swanage and the Isle of Purbeck

East Anglia:
- Frack Free East Anglia

Essex:
- Frack Free Essex

Gloucestershire:
- Frack Free Five Valleys

Greater Manchester:
- Barton Moss and Beyond
- Barton Moss Community Protection Camp
- Barton Moss Community Protection Camp Wishlist
- Barton Moss Protectors
- Bolton Against Fracking
- Daveyhulme and Trafford Frack Free
- Daveyhulme Community Protection Camp
- Frack Free Chorlton
- Frack Free Greater Manchester
- Frack Free Levenshulme
- Frack Free Salford & Manchester
- Frack Free Stockport
- Frack Free Tameside
- Frack Free Walkden
- Frack Free Wigan, Leigh and Makerfield
- IGas Barton Moss Drilling Site
- Irlam and Cattishead Frack Free
- Keep the Fracking Hell Out of Rochdale
- Say No to Fracking on Barton Moss

Hampshire and the Isle of Wight:
- East Hants Fracking Opposition Group
- Frack Free Isle of Wight
- Frack Free Rowlands Castle
- Frack Free Solent
- Isle of Wight - Fight Against Fracking!
- Solent Fracking Awareness

- Southampton Fracking Awareness

Herefordshire:
- Ban Fracking in Herefordshire

Kent:
- Adisham Against Fracking
- Bore Free Eastry
- Bromley Against Fracking
- East Kent Against Fracking
- Frack Free Darenth Valley
- Frack Free Kent
- Frack Free Sandwich
- Frack Free Whitstable
- Keep Shepherdswell Well
- No Fracking in Shepherdswell
- West Kent Against Fracking
- Whitstable Against Fracking

Lancashire:
- Blackpool & Fylde College Anti-Fracking Campaign
- Bowland Fracking Forum
- Bury Against Fracking
- Fleetwood Folk Say No Fracking (FFS No Fracking)
- Frack Free Blackpool
- Frack Free Burnley
- Frack Free Frek (Freckleton)
- Frack Free Fylde
- Frack Free Kirkham and Wesham
- Frack Free Lancashire
- Frack Free Lancashire Businesses
- Frack Free Lancashire – The Campaign
- Frack Free Pendle
- Frack Free Skelmersdale
- Frack Off Rossendale
- Garstang Against Fracking
- Hyndburn Anti Fracking Group
- Inskip Community
- Keep East Lancashire Frack Free
- Lancashire Community Protection Camp
- Lancaster Against Fracking
- Lancaster Fights Fracking
- Longridge Against Fracking
- Mereside United Frack Fighters

- Preston New Road Action Group
- Preston New Road (Westby with Plumpton) Against Fracking
- Ramsbottom Against Fracking
- Refracktion
- Residents Action on Fylde Fracking
- Ribble Estuary Against Fracking
- Ribchester Against Fracking
- Roseacre Awareness Group
- Rossendale Against Fracking
- Singleton Against a Fracked Environment
- The Nanas of Lancashire
- Thornton Unites Against Fylde Fracking

Leicestershire:
- Frack Free & Against Onshore Drilling Burton on the Wolds
- Frack Free & Against Onshore Drilling Leicestershire
- Frack Free Leicestershire

Lincolnshire:
- Frack Free Gainsborough
- Frack Free Lincolnshire

London:
- Frack Off London

Merseyside:
- Roby Residents Action Group – Against Fracking!
- Sefton Against Fracking

Nottinghamshire:
- Bassetlaw Against Fracking
- Daneshill Community Protection Camp
- Frack Free Nottinghamshire

Oxfordshire:
- Frack Free Wantage and Grove Oxfordshire
- Oxon Against Fracking

Shropshire:
- Dudleston Community Protection Camp
- Frack Free Dudleston
- Say No to Fracking in North Shropshire
- Shropshire Anti-Fracking

Somerset:
- Bruton Gasfield Free Community
- Frack Free Bristol
- Frack Free Chew Valley
- Frack Free Somerset
- Frack Free Yeovil and Surrounding Areas
- Fracking Awareness Backwell
- Gas Field Free Mendip
- Get the Frack Out Of The Mendips

Staffordshire:
- Gas Free Stoke
- No3Nooks Gas
- Tamworth Against Fracking

Surrey:
- Frack Free Horley
- Frack Free Surrey
- Horse Hill Protection Group

Sussex:
- Balcombe and Beyond
- Balcombe Community Protection Camp
- Brighton Action Against Fracking
- Chichester Anti-Fracking Forum
- Frack Free Arun
- Frack Free Balcombe
- Frack Free Balcombe Residents Association
- Frack Free Billingshurst
- Frack Free Fernhurst
- Frack Free FRow (Forest Row)
- Frack Free Horsham District
- Frack Free Markwells Wood
- Frack Free Mayfield and Five Ashes
- Frack Free Pulborough
- Frack Free Sussex
- Gas Drilling in Balcombe
- Gasfield Free Sussex
- Hastings & Rother Say NO to Fracking
- Heathfield Against Fracking
- Henfield Debates Fracking
- Horsham Against Fracking
- Keep Billingshurst Frack Free
- Keep Kirdford and Wisborough 'Green' (No Drilling)
- Knitting Nanas Balcombe
- Lewes Against Fracking
- No Fracking in Balcombe Society

- Sussex Extreme Energy Resistance
- Sussex Frack Free Food Alliance
- Warnham Against Fracking
- Wealden Against Fracking
- Worthing Against Fracking

Tyne and Wear:
- Frack Free Tyne & Wear

Warwickshire:
- Gasfield Free Leamington
- Gasfield Free Rugby
- No UCG Warks

West Midlands:
- Gasfield Free Birmingham
- Gasfield-free Coventry

Wiltshire:
- Keep Wiltshire Frack Free

Yorkshire:
- Crawberry Hill Community Protection Camp
- East Yorkshire Frack Free
- Frack Aware Barnsley
- Frack Free Doncaster
- Frack Free East Yorkshire
- Frack Free Harrogate and Knaresborough
- Frack Free Hebden Bridge
- Frack Free Hull & Holderness
- Frack Free Leeds
- Frack Free North Yorkshire
- Frack Free Ryedale
- Frack Free Scarborough
- Frack Free Skipton and Craven
- Frack Free South Yorkshire
- Frack Free York - Our Clean Energy Future
- Harrogate District Against Fracking
- Hull & East Yorkshire Against Extreme Energy
- Hull & East Yorkshire Against Fracking
- Hull & East Yorkshire Frack Off (HEY Frack Off)
- Keep Your Fracking Hands Off Calderdale
- The Battle for Crawberry Hill
- West Newton Community Protection Camp

- West Yorkshire Is NOT For Shale

Channel Islands
- No Fracking Jersey

Ireland
- Artists Against Fracking Northern Ireland
- Ballinlea Residents Group / Protect Our North Coast
- Ballymoney - Council Against Fracking
- Ballymoney Against Fracking
- Ban Fracking Northern Ireland
- Belcoo Frack Free
- Belfast Not For $hale
- Carrick Against Fracking
- Fermanagh Fracking Awareness Network
- Frack Events Ireland
- Frack Free Kerry
- Fracking Free Clare
- Fracking Free Ireland
- Lamp Fermanagh
- Love Leitrim
- No Fracking Dublin
- No Fracking Ireland
- No Fracking Leitrim
- No Fracking Northern Ireland
- No Fracking Sligo
- No Fracking Tyrone
- No Gas LAMP
- North West Network Against Fracking
- Protect Our Coast & Glens, Antrim
- Rathlin Bay Residents Say No to Fracking
- Stop Fracking Fermanagh

Scotland
- Anti-Fracking Renfrewshire
- Clacks Against Unconventional Gas Extraction
- Clydebank Anti Frack
- Clydebank Yes or No?
- Common Grounds, Edinburgh South
- Denny & Dunipace Against Unconventional Gas
- Don't Frack the Briggs (Bishopbriggs)
- Dunbar Anti-Fracking Team
- East Dunbartonshire Against Unconventional Gas Extraction
- Fight Against Fracking in Scotland
- Forth valley Against Unconventional Gas

- Frack Free Forth Valley
- Frack Off Ayrshire
- Frack Off Bearsden and Milngavie
- Frack Off Fife
- Frack Off Glasgow
- Frack Off Midlothian
- Frack Off Peeblesshire
- Frack Off Scotland
- Frack Off Stewarton & District
- Frack Off West Dunbartonshire
- Frack Off West Lothian
- Fracking Ban Scotland
- Halt Unconventional Gas Extraction, Cumberauld
- Hands Off Our Scotland
- Highlands & Islands Against Fracking
- No Fracking NB (North Berwick)
- No to Fracking in Scotland
- Our Forth – Portobello (Against Unconventional Gas)
- Scottish Pagans Against Fracking
- Scotland Against Fracking
- Shotts – Get Shot of Fracking & UGE
- South Lanarkshire Against (Unconventional) Gas
- West Dunbartonshire Against UCG
- West of Scotland Against Fracking

Wales
- Anti-Fracking West and Mid-Wales
- Borras & Holt Community Protection Camp
- Cardiff Against Fracking Everywhere
- Fracked Swansea
- Frack Free Deeside
- Frack Free Llantrisant
- Frack Free Newport
- Frack Free Porthcawl & Bridgend
- Frack Free Wales
- Frack Free Wrexham
- Frack Off Llanelli
- Frack (UCG) Free Swansea
- Keep Your Fracking Hands Off Neath Port Talbot
- Llantrithyd Villagers Against Fracking
- No Fracking Way! Beacons Anti-Fracking Group

Appendix One

- North East Wales Anti-Fracking Action
- North East Wales Anti-Fracking Network
- Say No to Fracking in South Wales
- Say No to Toxic Gas Drilling in the Vale
- South Wales Against Fracking Festival
- Swansea Against Fracked Energy (SAFE)
- The Vale says NO!

APPENDIX TWO

GLOBAL FRACKING BANS

Below is a list (correct as of December 2014) of all known fracking bans and moratoria (including Coal Seam Gas) around the world. It also includes a small number of bans that have been overturned or rescinded. Many of the bans are official (such as those in the USA and nationwide moratoria), but unofficial bans, such as resolutions passed by regional councils opposing unconventional drilling are also included. In the United Kingdom, outrights bans are not possible under current planning and political systems. Nonetheless, several Councils have declared themselves to be 'Frack Free Zones' or recorded Statements of Opposition and have pledged to refuse permission for any applications for unconventional drilling. A few Councils have noted Statements of Concern at fracking, but have not necessarily pledged to refuse an application.

Argentina
- Cinco Saltos, Patagonia
- Five City Breaks, Black River Province

Australia
- Dunoon Community
- Modanville Community
- New South Wales State (rescinded)
- Numulgi Community
- Poowong Community
- Rosebank Community
- The Channon Community
- Tyalgum Community
- Victoria State (temporary)
- Whian Whian Community

Bulgaria
Whole Nation

Canada
- First Nations Community
- Nova Scotia
- Quebec

Czech Republic (temporary)
Whole Nation

France
Whole Nation

Italy
Whole Nation

Germany
Whole Nation

Luxemburg
Whole Nation

Netherlands
- Aalburg Municipality
- Aalten Municipality
- Achtkarspelen Municipality
- Almelo Municipality
- Alphen aan den Rijn Municipality
- Alphen-Chaam Municipality
- Amersfoort Municipality
- Amsterdam Municipality
- Amsterdam-Zuidoost Borough
- Apeldoorn Municipality
- Arnhem Municipality
- Asten Municipality
- Baarle-Nassau Municipality
- Beesel Municipality
- Bellingwedde Municipality
- Bergeijk Municipality
- Bergen Municipality
- Bergen op Zoom Municipality
- Berkelland Municipality
- Bernisse Municipality
- Best Municipality
- Beuningen Municipality
- Binnenmaas Municipality
- Bladel Municipality
- Blaricum Municipality
- Bloemendaal Municipality
- Bodegraven-Reeuwijk Municipality
- Boskoop Municipality (now part of Alphen aan den Rijn Municipality)
- Boxmeer Municipality
- Boxtel Municipality
- Breda Municipality
- Bronckhorst Municipality
- Brummen Municipality
- Bunnik Municipality
- Buren Municipality
- Bussum Municipality
- Castricum Municipality
- Cuijk Municipality
- Culemborg Municipality
- Dalfsen Municipality
- Dantumadiel Municipality
- De Bilt Municipality
- De Ronde Venen Municipality
- Delfzijl Municipality
- Den Haag Municipality

- Diemen Municipality
- Dinkelland Municipality
- Doesburg Municipality
- Doetinchem Municipality
- Dongen Municipality
- Dongeradeel Municipality
- Dordrecht Municipality
- Drimmelen Municipality
- Duiven Municipality
- Ede Municipality
- Eemsmond Municipality
- Eersel Municipality
- Eijsden-Margraten Municipality
- Eindhoven Municipality
- Enschede Municipality
- Epe Municipality
- Ermelo Municipality
- Ferwerderadiel Municipality
- Flevoland Province
- Friesland Province
- Gaasterlân-Sleat Municipality (now part of De Friese Meren Municipality)
- Geertuidenberg Municipality
- Gelderland Province
- Geldermalsen Municipality
- Geldrop-Mierlo Municipality
- Gennep Municipality
- Gilze en Rijen Municipality
- Goeree-Overflakkee Municipality
- Goirle Municipality
- Gorinchem Municipality
- Grave Municipality
- Groesbeek Municipality
- Groningen Province
- Grootegast Municipality
- Gulpen-Wittem Municipality
- Haaksbergen Municipality
- Haaren Municipality
- Haarlem Municipality
- Haarlemmerliede en Spaarnwoude Municipality
- Haarlemmermeer Municipality
- Halderberge Municipality
- Hardenberg Municipality
- Heemstede Municipality
- Heerde Municipality
- Heerenveen Municipality
- Heerlen Municipality
- Heeze-Leende Municipality
- Hellendoorn Municipality
- Helmond Municipality
- Hengelo Municipality
- Heumen Municipality
- Heusden Municipality
- Hilvarenbeek Municipality
- Hilversum Municipality

- Hof van Twente Municipality
- Hoogezand-Sappemeer Municipality
- Huizen Municipality
- IJsselstein Municipality
- Kampen Municipality
- Kerkrade Municipality
- Kollumerland en Nieuwkruisland Municipality
- Landgraaf Municipality
- Lansingerland Municipality
- Laren Municipality
- Leek Municipality
- Leiden Municipality
- Leidschenveen-Ypenburg District
- Lemsterland Municipality (now part of De Friese Meren Municipality)
- Leudal Municipality
- Limburg Province
- Lingewaard Municipality
- Littenseradiel Municipality
- Lochem Municipality
- Loon op Zand Municipality
- Loppersum Municipality
- Losser Municipality
- Maasdonk Municipality
- Maassluis Municipality
- Maastricht Municipality
- Marum Municipality
- Meerssen Municipality
- Midden-Delfland Municipality
- Mill en Sint Hubert Municipality
- Millingen aan de Rijn Municipality
- Moerdijk Municipality
- Montferland Municipality
- Mook en Middelaar Municipality
- Muiden Municipality
- Naarden Municipality
- Neder-Betuwe Municipality
- Nieuwegein Municipality
- Nieuwkoop Municipality
- Nijmegen Municipality
- Noord-Brabant Province
- Noordenveld Municipality
- Noord-Holland Province
- Noordoostpolder Municipality
- Noordwijk Municipality
- Oirschot Municipality
- Oisterwijk Municipality
- Oldenzaal Municipality
- Olst-Wijhe Municipality
- Oost Gelre Municipality
- Oosterhout Municipality
- Ooststellingwerf Municipality
- Opsterland Municipality
- Oss Municipality

- Oude IJsselstreek Municipality
- Oudewater Municipality
- Overijssel Province
- Papendrecht Municipality
- Pekela Municipality
- Purmerend Municipality
- Reusel-De Mierden Municipality
- Rheden Municipality
- Ridderkerk Municipality
- Rijssen-Holten Municipality
- Roerdalen Municipality
- Roermond Municipality
- Rozendaal Municipality
- Schijndel Municipality
- Schoonhoven Municipality
- Schouwen-Duiveland Municipality
- Sint-Michielsgestel Municipality
- Sint-Oedenrode Municipality
- Sittard-Geleen Municipality
- Skarsterlân Municipality (now part of De Friese Meren Municipality)
- Sliedrecht Municipality
- Slochteren Municipality
- Smallingerland Municipality
- Someren Municipality
- Son en Breugel Municipality
- Stadskanaal Municipality
- Steenbergen Municipality
- Steenwijkerland Municipality
- Tholen Municipality
- Tiel Municipality
- Tilburg Municipality
- Tubbergen Municipality
- Twenterand Municipality
- Tytsjerksteradiel Municipality
- Ubbergen Municipality
- Urk Municipality
- Utrecht Municipality
- Utrecht Province
- Utrechtse Heuvelrug Municipality
- Veendam Municipality
- Veenendaal Municipality
- Veghel Municipality
- Veldhoven Municipality
- Venlo Municipality
- Vianen Municipality
- Vlaardingen Municipality
- Vlagtwedde Municipality
- Voerendaal Municipality
- Voorst Municipality
- Vught Municipality
- Waalre Municipality
- Waalwijk Municipality
- Wageningen Municipality
- Waterland Municipality
- Weert Municipality
- Weesp Municipality

- Westerveld Municipality
- Weststellingwerf Municipality
- Westvoorne Municipality
- Wierden Municipality
- Wijchen Municipality
- Wijdemeren Municipality
- Wijk bij Duurstede Municipality
- Winsum Municipality
- Winterswijk Municipality
- Woensdrecht Municipality
- Woerden Municipality
- Wormerland Municipality
- Woudrichem Municipality
- Zaanstad Municipality
- Zaltbommel Municipality
- Zeist Municipality
- Zevenaar Municipality
- Zuidhorn Municipality
- Zundert Municipality
- Zutphen Municipality
- Zwartewaterland Municipality
- Zwijndrecht Municipality
- Zwolle Municipality

New Zealand
- Christchurch City
- Kaikoura District

Republic of Ireland
- Whole Nation (temporary)
- County Leitrim
- County Clare
- Fingal County
- Roscommon County
- Sonegal and Sligo District

Romania (rescinded)
Whole Nation

South Africa (rescinded)
Whole Nation

Spain
- Cantabria
- Fuerteventura Biosphere Reserve
- Valle de Mena

Switzerland
- Canton of Fribourg

United Kingdom
- Bath and North East Somerset District[3]
- Brent Borough of London[2]
- Brighton and Hove City[1]
- Bristol City[2]
- Calderdale Borough[1]
- Cheshire East District[2]
- Cheshire West and Chester District[2]

- Deal Town[2]
- Dover District[2]
- East Sussex[3]
- Fermanagh District[1]
- Glastonbury Town[2]
- Hampshire[3]
- Hull City[3]
- Inverclyde[3]
- Kirklees Borough[3]
- London[1]
- Manchester City[1]
- Mendip District[3]
- Newcastle-upon-Tyne City[2]
- North Lanarkshire[2]
- Pendle Borough[3]
- Preston City[1]
- Sheffield City[2]
- Trafford City[1]
- Waltham Forest Borough of London[2]
- Westhoughton Town[2]
- Wirral Metropolitan Borough[2]
- York City[1]

[1] Frack Free Zone
[2] Statement of Opposition
[3] Statement of Concern

USA
California:
- Arroyo Grande
- Berkeley
- Beverly Hills
- Cambria Community Services District
- Carson
- Compton
- Culver City
- Fairfax
- Los Angeles
- Los Angeles Community College District
- Mar Vista Community Council, Los Angeles
- Marin County
- Oakland
- Rampart Village Neighbourhood Council, Los Angeles
- San Luis Obispo
- Santa Cruz County
- San Francisco
- Sebastopol
- Sonoma

Colorado:
- Boulder City
- Boulder County
- Broomfield
- Colorado Springs
- Erie
- Fort Collins
- Lafayette
- Longmont
- Loveland
- Nederland

Connecticut:
State of Connecticut

District of Columbia:
District of Columbia

Florida:
- Coconut Creek
- Hallandale Beach

Hawaii:
- Hawai'i County

Illinois:
- Alto Pass
- Anna
- Carbondale
- Jackson County
- Johnson County
- Murphysboro
- Pope County

Indiana:
- Terre Haute

Iowa:
- Allamakee Country

Maryland:
- Baltimore
- College Park
- Mountain Lake Park
- Washington Grove

Massachusetts:
- Chesterfield

Michigan:
- Burleigh Township
- Cannon Township
- Courtland Township
- Cross Village Township
- Dearborn Heights
- Detroit
- Ferndale
- Heath Township
- Ingham County
- Orangeville Township
- Reno Township
- Scio Township
- Shelby Township
- Southfield
- Thornapple Township
- Waterford Township
- Wayne County
- West Bloomfield Township
- Yankee Springs Township
- Ypsilanti

Minnesota:
- Goodhue County
- Houston County

New Jersey:
- State of New Jersey
- Bethlehem
- Bloomingdale
- Bordentown Township
- Byram Township
- Clinton
- Clinton Township
- Closter
- Delaware Township

- Franklin Township, Somerset County
- Glassboro
- Greenwich Township
- Highland Park
- Hillsdale
- Holland
- Howell Township
- Lambertville
- Lebanon Township
- Middlesex County
- Monmouth Beach
- Musconetcong River Management Council
- New Brunswick
- Paramus
- Passaic County
- Point Pleasant
- Princeton Township
- Princeton Borough
- Readington
- Red Bank
- Rumson
- Rutgers University Student Assembly
- Secaucus
- Stillwater
- Stockton
- Trenton
- Union Beach

New Mexico:
- Las Vegas
- Mora County
- San Miguel County

New York:
- State of New York
- Albany
- Albany County
- Alfred Town
- Alfred Village
- Altamont
- Amherst
- Andes
- Annsville
- Auburn
- Augusta
- Ava
- Avon Town
- Barrington
- Beacon
- Benton
- Berne
- Bethel
- Binghamton (overturned)
- Blenheim
- Boonville Town
- Buffalo
- Butternuts
- Brighton
- Bristol
- Brookfield
- Caledonia Town
- Camden
- Camillus
- Canandaigua
- Canandaigua Lake Watershed Association
- Canandaigua Town
- Carmel
- Caroline

- Cayuga County
- Cherry Valley
- Chester Town, Orange County
- Clinton County
- Cochecton
- Colden
- Conesus
- Cooperstown Chamber of Commerce
- Cooperstown
- Copake
- Cortland
- Cortland County
- Cortlandt
- Cortlandville
- Danby
- Danube
- Deerfield
- DeWitt
- Dolgeville
- Dryden
- Dunkirk
- Eaton
- Elbridge
- Enfield
- Erie County
- Fabius
- Floyd
- Forestburgh
- Forestport
- Freeville
- Fulton Town
- Geneseo Town
- Geneva
- Geneva Town
- Genoa
- Germantown
- Gorham
- Guilderland
- Hartwick
- Highland
- Hopewell
- Hudson
- Huron
- Italy
- Ithaca City
- Ithaca Town
- Jerusalem
- Kirkland
- LaFayette
- Lancaster Town
- Lansing
- Lebanon
- Ledyard
- Lenox
- Lima Town
- Lincoln
- Little Falls
- Livonia
- Locke
- Lumberland
- Manchester
- Manheim
- Marbletown
- Marshall
- Marcellus
- Mendon
- Meredith
- Middleburgh Town
- Middlefield
- Middlesex

- Milford Town
- Milo Town
- Minden
- Moravia
- Naples Town
- Naples Village
- Nassau County
- Newfield
- New Hartford
- New Lisbon
- New Paltz Town
- New Paltz Village
- Newport
- New York City
- Niagara Falls
- Niles
- Niskayuna
- Nunda Town
- Olean
- Olive
- Oneida County
- Oneonta
- Oneonta Town
- Onondaga County
- Onondaga Town
- Ontario County
- Oppenheim
- Orange County
- Otego Town
- Otisco Town
- Otsego Town
- Owasco
- Owego Village
- Oxford Village
- Palatine
- Paris
- Penfield
- Penn Yan
- Perinton
- Phelps
- Philipstown
- Plainfeld
- Plattsburgh
- Pompey
- Portage Town
- Preble
- Putnam County
- Red Hook
- Remsen
- Rensselaerville
- Richmond
- Richmondville Town
- Rochester City
- Rochester Town
- Rockland County
- Rome
- Roseboom
- Rosendale
- Rush
- Rushford Lake Recreation District
- Sangerfield
- Saugerties
- Schoharie Town
- Scipio
- Seward
- Sennett
- Sharon
- Sharon Springs
- Skaneateles
- South Bristol
- Southeast

- Southampton Town
- Spafford
- Springfield
- Springwater
- St. Johnsville Town
- St. Johnsville Village
- Stafford
- Starkey
- Suffolk County
- Sullivan County
- Summerhill
- Syracuse
- Taghkanic
- Tompkins County
- Torrey
- Trenton
- Trumansburg
- Tully
- Tusten
- Ulster County
- Utica
- Ulysses
- Vienna
- Virgil
- Vernon
- Verona
- Wales
- Warwick
- Waterloo Town
- Wawarsing
- Wayland Town
- Wayne
- West Bloomfield
- Westchester County
- Westerlo
- Westmoreland
- West Sparta
- Whitesboro
- Whitestown
- Wilson Village
- Woodstock
- Yates County
- Yorkshire

North Carolina:
- Anson County
- Bertie County
- Butner
- Camden County
- Carrboro
- Carteret County
- Chapel Hill
- Creedmoor
- Currituck County
- Dare County
- Duck
- Durham
- Durham County
- Elizabeth City
- Granville County
- Greenville
- Haywood County
- Jones County
- Kill Devil Hills
- Orange County
- Pasquotank County
- Perquimans County
- Pittsboro
- Raleigh
- Southern Shores
- Stokes County
- Sunset Beach

- Tyrrell County

Ohio:
- Amesville
- Athens
- Athens County
- Bowling Green
- Broadview Heights
- Brunswick
- Burton
- Canal Fulton
- Canton
- Chester Township, Geauga County
- Cincinnati
- Columbiana
- Garrettsville
- Girard
- Hartville
- Heath
- Hinckley Township
- Lake Erie
- Madison Township, Richland County
- Mansfield
- Medina Township
- Meyers Lake
- Montville Township, Medina County
- Munroe Falls
- Niles
- North Canton
- Oberlin
- Plain Township
- Randolph Township
- Sharon Township, Medina County
- South Russell
- Stow
- Summit County
- Weathersfield Township
- Yellow Springs
- Youngstown
- York Township, Medina County

Pennsylvania:
- Baldwin
- Buckingham Township Civic Association
- Easton
- Emsworth
- Ferguson Township
- Forest Hills
- Harveys Lake
- Lanesboro
- Media Borough
- Murrysville
- New Hope
- Philadelphia
- Phoenixville
- Pittsburgh
- State College Borough
- West Homestead
- Wilkinsburg

Texas:
- Bartonville
- Denton
- Dish
- Flower Mound

Vermont:
State of Vermont

Virginia:
- Bath County
- Botetourt County
- Fairfax County Water Authority
- Falls Church
- Headwaters Soil & Water Conservation District
- Lynchburg
- Roanoke
- Rockbridge County
- Shenandoah County
- Staunton

West Virginia:
- Lewisburg
- Morgantown (overturned)
- Pocahontas County Free Libraries
- Wellsburg (repealed)

Wisconsin:
- Eau Claire County
- Ellsworth Town
- Holland, La Crosse County
- Oak Grove, Pierce County
- Rock Elm

Wyoming:
- Bridger-Teton National Forest

Native American Communities:
- Haudenosaunee Environmental Task Force
- Turtle Mountain Band of Chippewa

APPENDIX THREE

NORTH AMERICAN FRACKING INCIDENTS JANUARY 2013 TO DECEMBER 2014

Below is a partial list of confirmed incidents related to hydraulic fracturing in the USA and Canada in the period January 2013 to December 2014 as reported in various media outlets. This data is reproduced with the kind permission of the members of the Marcellus Outreach Butler group, based in Pennsylvania, USA.

The data is listed in date order and the incidents categorized into one of six types as follows:

Blowout – eruption of oil, fluids and other material from a wellhead as a result of malfunction or pressure build up.

Explosion – the violent detonation of vehicles, trains, wells and other facilities.

Pipeline/Infrastructure – incidents related to pipelines or other infrastructure except for explosions.

Spill/Dump – cases of accidental spillages and willful dumping of fluids and/or materials.

Traffic – incidents involving vehicles connected to hydraulic fracturing operations such as road traffic collisions.

Other – incidents that do not fit into the above categories.

Date	Incident Type	Detail
1 November 2012 to 31 January 2013	Spill/Dump	More than 250,000 gallons (946,353 litres) of waste fluids and oil illegally dumped into a river in Ohio.
14 January 2013	Explosion	Two workers critically injured following an explosion at a Texas well site.
16 January 2013	Other	Chemical emergency reported at an Ohio oil well. No inventory of chemicals used on site was available to local authorities at the time of investigation.
4 February 2013	Spill/Dump	Around 840 gallons (3,180 litres) of waste fluids accidentally spilled in Pennsylvania.
11 February 2013	Spill/Dump	84,000 gallons (317,975 litres) of fracking fluids spilled from a well over a period of nearly 30 hours.
12 February 2013	Spill/Dump	12,000 gallons (45,425 litres) of fluids spilled at a Pennsylvanian well site.
13 February 2013	Blowout	A blowout at a well in Chesapeake polluted a nearby stream with large quantity of fluid.
22 February 2013	Spill/Dump	More than 95,000 gallons (359,614 litres) of fluids stored in an open air pond spilled into a local stream in West Virginia.
26 February 2013	Other	One worker was killed and another injured in a drilling accident in Ohio.
March 2013	Spill/Dump	Petroleum-based fracking fluids found illegally dumped into a creek. The date of the offence is not known.
9 March 2013	Pipeline/Infrastructure	A compressor station in Pennsylvania released methane and other pollutants for three hours.

9 March 2013	Traffic	Two children killed when a water tanker rolled onto a car in West Virginia.
14 March 2013	Spill/Dump	Fracking fluid leaked from a Pennsylvanian well site at a rate of 800 gallons (3,028 litres) per minute.
15 March 2013	Explosion	A gas well near Chippewa Township in Ohio exploded sending flames 30 feet (9.1m) into the air and caused the evacuation of the town.
17 March 2013	Explosion	An oil tank on an Ohio well pad exploded, sending debris over 400 feet (122m) into the garden of a nearby house.
19 March 2013	Pipeline/Infrastructure	A compressor station in Pennsylvania caught fire and injured one worker.
4 April 2013	Explosion	A compressor station at an Oklahoma gas facility exploded and forced the evacuation of all homes within one square mile (2.6 square km).
11 April 2013	Explosion	Two men killed in a pipeline explosion in West Virginia.
24 April 2013	Traffic	A 14-year-old boy was crushed to death after being hit by an 18-wheeler tanker carrying fracking waste. The driver failed to stop but was later cleared of any wrong doing.
27 April 2013	Spill/Dump	More than 100 barrels (15,899 litres) of oil-based mud spilled into a creek following a traffic accident in West Virginia.
27 April 2013	Pipeline/Infrastructure	A large quantity of gas and oil residue was expelled during a pipeline blowout near Lafayette Township.

30 April 2013	Spill/Dump	9,000 gallons (34,069 litres) of waste fluid spilled onto a miniature horse farm, flooding the farmhouse basement and garage in Wyoming County.
1 May 2013	Spill/Dump	More than 1,600 gallons (6,057 litres) of oil spilled from a storage tank into a creek in Ohio.
14 May 2013	Explosion	A natural gas compressor in Pennsylvania exploded.
22 May 2013	Other	A worker was killed at a drilling site in North Dakota.
30 May 2013	Explosion	Thirteen people were injured in a pipeline explosion in New Jersey.
30 May 2013	Spill/Dump	A quantity of pollutants entered a watercourse during drilling operations in West Virginia.
1 June 2013	Spill/Dump	A call to the National Report Center reported a tanker illegally dumping fluids into a stream in West Virginia.
1 June 2013	Pipeline/Infrastructure	A pipeline in Canada leaked 2.5 million gallons (9.5 million litres) of waste fluids.
4 June 2013	Other	A haulage company in Ohio was ordered to suspend operations in that State after evidence was released linking the company to illegal dumping of waste into a private pond.
4 June 2013	Pipeline/Infrastructure	A pipeline construction company was fined $150,000 for multiple incidents of pollution during the construction of a pipeline in Pennsylvania.
7 June 2013	Spill/Dump	A drilling company was found guilty of illegally disposing of produced and fracking fluids into an unlined pit in California.
12 June 2013	Explosion	A facility in Los Angeles exploded, killing two people and injuring 100 others.

12 June 2013	Explosion	A lorry accidentally hit a high-pressure line and caused an explosion.
13 June 2013	Pipeline/Infrastructure	A gas pipeline underneath the Ohio River ruptured and released large quantities of gas into the river between Ohio and West Virginia.
18 June 2013	Explosion	A Pipeline in Louisiana ruptured, causing an explosion and fire.
20 June 2013	Pipeline/Infrastructure	A gas pipeline containing toxic hydrogen sulphide gas (amongst others) ruptured and forced the evacuation of 50 homes after being hit by debris from a flood.
21 June 2013	Explosion	A worker died from burns sustained in an explosion at a well pad.
21 June 2013	Other	A Glycol Dehydration plant accidentally discharged pollutants into the atmosphere in the 3rd such incident since 2012.
22 June 2013	Traffic	A water tanker failed to observe a stop sign and killed a mother and her 14-year-old daughter.
25 June 2013	Spill/Dump	A well pad in Pennsylvania spilled around 20 barrels (3,180 litres) of oil onto a well pad after a valve was left open. The oil leached into the ground through a hole in the impermeable membrane.
25 June 2013	Explosion	A pipeline exploded in Los Angeles forcing 55 people to be evacuated from nearby properties.
25 June 2013	Other	A call to the National Report Center reported a fire at an Ohio oil well. 150 gallons (568 litres) of oil was released into the environment.

26 June 2013	Pipeline/Infrastructure	Two people were slightly injured and 34 cars damaged when a section of road suddenly subsided due to a pipeline project.
6 July 2013	Explosion	A massive explosion caused by a derailed oil train destroyed parts of Lac-Megantic in Quebec, Canada, killing as many as 40 people and forcing 1,000 more to be evacuated from their homes.
7 July 2013	Explosion	Up to eight people were injured when a West Virginian gas well exploded.
8 July 2013	Spill/Dump	Fracking and produced fluids were spilled at a well in Lycoming County.
8 July 2013	Pipeline/Infrastructure	An upgrade project to a pipeline caused a road in New Jersey to suddenly collapse. It took over a week to repair the road.
8 July 2013	Spill/Dump	A gasfield worker admitted to illegally dumping waste fluids into a river after allegedly being instructed to do so.
9 July 2013	Other	Two workers were killed at an oilfield accident in Kansas after being exposed to toxic hydrogen sulphide gas.
9 July 2013	Other	A gas leak in the Gulf of Mexico released a large quantity of methane into the atmosphere. The Coast Guard reported observing a rainbow sheen over four square miles (10.4 square km) of the sea.
12 July 2013	Spill/Dump	An accidental spill of fluids on a well site killed vegetation along a large tract of land.
13 July 2013	Other	A gas well in West Virginia caught fire.

14 to 15 July 2013	Other	An incident at a Washington County gas plant sent large quantities of black smoke into the air for two days. Nearby residents reported an audible 'boom' before the smoke cloud.
16 July 2013	Spill/Dump	A drilling company was found guilty of illegally discharging industrial waste into watercourses in Clearfield County.
18 July 2013	Spill/Dump	A wastewater treatment company received a legal notice from Clean Water Action for an alleged illegal discharge of toxic waste fluids into a river in Pennsylvania.
18 July 2013	Spill/Dump	A company was fined $100,000 for discharging up to 57,373 gallons (217,180 litres) of waste fluids into a river in Pennsylvania over a two month period.
18 July 2013	Spill/Dump	An Ohio landowner alleged that a drilling company illegally dumped waste fluids on his land.
18 July 2013	Spill/Dump	A drilling company is found guilty of illegally discharging waste fluids into watercourses in Pennsylvania.
20 July 2013	Traffic	A driver of a water tanker was airlifted to hospital after his vehicle overturned in West Virginia.
21 July 2013	Traffic	A tanker leaked hydrochloric acid onto the road, causing a section of Interstate 70 to close for four hours.
22 July 2013	Traffic	A crane rolled of a road in West Virginia and leaked diesel, causing the road to close.
22 July 2013	Pipeline/Infrastructure	Several hundred people were evacuated from their homes in Ohio after a high-pressure gas pipeline ruptured.

Date	Type	Description
23 July 2013	Other	A gas well off the Louisiana Coast caught fire after operators lost control of the rig.
24 July 2013	Explosion	A worker was killed in an explosion at a West Virginian well pad.
24 July 2013	Explosion	A tanker carrying waste fracking fluid exploded in Texas, killing a nearby resident.
25 July 2013	Other	The American Justice Department announces the admission by Halliburton to destroying evidence relating to the Deepwater Horizon disaster in the Gulf of Mexico and will plead guilty to a criminal charge.
25 July 2013	Pipeline/Infrastructure	Waste from the construction of a pipeline leaked into a stream in Ohio.
28 July 2013	Explosion	A worker was killed following an explosion at a well pad in West Virginia.
August 2013	Blowout	A North Dakotan oil well suffered its 11th blowout since 2006 and sent 655,823 litres (173,250 gallons) of contaminants into the air. The previous 10 blowouts had released a combined total of 435,322 litres (115,000 gallons).
1 August 2013	Traffic	A fracking fluid tanker collided with two cars in Ohio, causing injuries.
1 August 2013	Traffic	A tanker carrying ethylene glycol crashed and spilled 265 gallons (1,003 litres) of the contents onto a nearby field.
3 August 2013	Spill/Dump	Five barrels (795 litres) of produced fluids leaked from a hole in the coil tubing unit and spilled onto the well pad. Further holes were found in the impermeable membrane.

7 August 2013	Spill/Dump	An official study into a pollution incident in Kentucky in 2007 concluded that a fracking waste spill was the likely cause of the widespread death of fish.
10 August 2013	Other	A worker was killed after being hit by a falling block at an Alberta well site.
13 August 2013	Explosion	Around 80 families were forced to evacuate their homes after a nearby pipeline ruptured, caught fire and exploded in unknown circumstances. It was carrying ethane and propane gas.
13 August 2013	Pipeline/Infrastructure	A landslip ruptured a pipeline in West Virginia, causing it to leak an unknown liquid into a stream.
16 August 2013	Other	A West Virginian well site caught fire and injured three workers.
16 August 2013	Pipeline/Infrastructure	A cow was killed after being exposed to hydrogen sulphide leaking from a pipeline in Alberta.
20 August 2013	Pipeline/Infrastructure	A drilling company was found guilty of causing the deaths of a number of fish after a pipeline ruptured and spilled its contents into the water.
20 August 2013	Traffic	A lorry working for oil and gas wells in West Virginia hit and killed a 57-year-old man.
20 August 2013	Explosion	A gas pipeline exploded in Oklahoma, sending flames 200 feet (61m) into the air and damaging a nearby barn.
21 August 2013	Spill/Dump	A drilling company was found guilty of not reporting two cases of accidental spillages in Pennsylvania after soil samples and eye-witnesses suggested waste fluids spilled onto the ground.

22 August 2013	Spill/Dump	5,000 gallons (18,927 litres) of sulphuric acid was spilled into a river in the Marcellus Shale region.
22 August 2013	Other	A number of nearby residents complained of burning eyes, nausea and other symptoms following a flaring incident at a gas processing plant in Pennsylvania.
23 August 2013	Spill/Dump	A breach in a fracking fluid storage pond flooded a nearby house in Texas.
27 August 2013	Spill/Dump	A call to the National Report Center reported a tanker illegally dumping fracking fluid along a roadway and into a stream in Ohio.
29 August 2013	Pipeline/Infrastructure	A worker was killed in a pipeline construction accident in Ohio. A report was filed by the worker prior to the accident indicating a fault with the equipment.
29 August 2013	Explosion	An oil rig exploded in Texas, causing a massive fire.
29 August 2013	Spill/Dump	A worker admits in a Cleveland Court that he was instructed to illegally dispose of waste fracking fluids into the nearby domestic sewer system on 24 separate occasions.
30 August 2013	Pipeline/Infrastructure	A pipeline leaked an unknown amount of natural gas in the Gulf of Mexico.
September 2013	Blowout	More than 200 barrels (31,797 litres) of fracking fluid, oil and waste fluids erupted out of a conventional oil well in North Montana as a result of pressure from a nearby fracking operation. The fluids were expelled half a mile (0.8km) across the ground.

10 September 2013	Spill/Dump	The State of Pennsylvania filed criminal charges against a drilling company for illegally dumping more than 50,000 gallons (189,271 litres) of toxic waste from a fracking well in Lycoming County.
11 September 2013	Other	A 21-year-old-worker was killed on a fracking well pad in North Dakota.
11 to 12 September 2013	Spill/Dump	Widespread flooding in Colorado results in 11 'notable' oil spills amounting to 34,524 gallons (130,688 litres) of oil.
16 September 2013	Explosion	A chemical plant in Oklahoma exploded and burned for over nine hours.
21 September 2013	Explosion	A gas processing plant in West Virginia exploded in the early hours, causing nearby residents to evacuate.
21 September 2013	Explosion	A gas pipeline in Texas suddenly ruptured and exploded.
21 September 2013	Other	A gas processing plant in West Virginia caught fire and burned for around eight hours. Eleven homes were evacuated.
25 September 2013	Spill/Dump	20 gallons (76 litres) of flowback fluid spilled and leached into the ground at a well site in Lycoming County, Pennsylvania.
27 September 2013	Traffic	A tanker caught fire and exploded at an Ohio well site during a transfer of fluids.
28 September 2013	Explosion	A high-pressure gas well exploded, injuring a worker.
28 September 2013	Explosion	Five storage containers at a waste disposal plant in Texas exploded after being hit by lightning.

29 September 2013	Pipeline/Infrastructure	A farmer discovered a break in a pipeline spilled 20,600 barrels (3.27 million litres) of oil from fracking operations over a wheat field. The cleanup was estimated as costing $4 million and taking several years.
1 October 2013	Pipeline/Infrastructure	Drilling operations for a pipeline construction in Ohio damaged two properties and contaminated water.
1 October 2013	Traffic	A collision involving two tankers and another vehicle resulted in one injury.
1 October 2013	Other	A well pad in Westmoreland County was found to be unsound in construction, having a torn impermeable membrane, ground spillages and unlawfully adding Portland cement to drill cuttings.
2 October 2013	Other	A worker admits to unlawfully injecting waste fluids into a well despite regulators advising the well was not in an acceptable condition.
2 October 2013	Spill/Dump	A University study discovered radium in concentrations 200 times higher than normal levels and slats such as bromide in a creek in Pennsylvania.
2 October 2013	Traffic	A tanker driver was killed after leaving the road and rolling into a ditch. Fluids were also observed leaking from the upturned vehicle.
3 October 2013	Other	An injection well in Oklahoma was shutdown after a swathe of earthquakes in the vicinity.
7 October 2013	Explosion	A worker was injured following an explosion at a waste disposal site in Texas.

8 October 2013	Explosion	At least one person injured after a waste disposal well exploded. Nearby roads were forced to close.
9 October 2013	Traffic	A tanker carrying fracking fluids got stuck in an underpass and overturned in Ohio.
11 October 2013	Traffic	A lorry transporting drill pipes crashed in Texas.
25 October 2013	Spills/Dumps	Associated Press announces records identifying 750 oilfield incidents, including 300 pipeline spillages in North Dakota since January 2012 that were not publically reported.
27 October 2013	Other	Two workers were burned at a gas well in Doddridge County.
29 October 2013	Pipeline/Infrastructure	17,000 gallons (64,352 litres) of oil leaked from a pipeline in Texas.
7 November 2013	Explosion	A saltwater tank exploded at a well site in North Dakota, spilling 2,700 barrels (429,266 litres) of waste fluid.
14 November 2013	Explosion	A gas pipeline exploded half a mile (0.8km) from a school in Ellis County. Families were forced to evacuate within a 3 mile (4.8km) radius for up to 36 hours.
25 November 2013	Pipeline/Infrastructure	17,000 barrels (2.7 million litres) of fracking fluid leaked into a mile (1.6km) long stretch of a stream in North Dakota.
29 November 2013	Traffic	A tanker carrying fluids from a gas well site rolled into a creek near a Methodist Church in West Virginia. The driver was injured.
29 November 2013	Explosion	A pipeline in Montana exploded in a huge fireball that could be seen for 30 miles. (48km) Three homes were evacuated.

Date	Type	Description
5 December 2013	Pipeline/Infrastructure	Workers accidentally vent millions of gallons of gas from a ten mile (16km) long section of pipeline, forcing a local family to evacuate their home.
6 December 2013	Explosion	A gas storage tank on a Texas well site exploded.
9 December 2013	Spill/Dump	Waste fluid spilled onto a well pad in Pennsylvania.
11 December 2013	Spill/Dump	The freezing weather caused a line on a drill site to rupture.
16 December 2013	Traffic	A tanker carrying hexamine en route to a drilling site in West Virginia spilled its contents onto the road, closing all four lanes.
21 December 2013	Other	A fire at a gas well in Texas is suspected to have caused by a lightning strike.
23 December 2013	Traffic	A 60-year-old pedestrian was hit and killed by a tanker in Pennsylvania. The driver and passenger were drunk and fled the scene on foot.
27 December 2013	Spill/Dump	Waste was spilled at a drilling site in the Allegheny National Forest.
30 December 2013	Explosion	An oil-carrying train crashed and exploded into a fire that burned for more than 24 hours in North Dakota.
2 January 2014	Explosion	A storage tank at a West Virginian well site exploded, injuring one worker and spreading fluids on the ground.
6 January 2014	Pipeline/Infrastructure	A gas compressor station in New York State caught fire.
7 January 2014	Explosion	A train carrying oil and propane gas derailed and exploded in New Brunswick, causing 150 residents to evacuate their homes.

Date	Type	Description
9 January 2014	Other	A worker is airlifted to hospital after sustaining a head injury whilst clearing ice from a pipeline in Ohio.
16 January 2014	Other	A Texas family blame nearby injection wells for a spate of 32 earthquakes that had damaged their property.
18 January 2014	Spill/Dump	Open air waste pits flood after heavy rain in Canada.
20 January 2014	Other	A train carrying oil and sand derailed on a bridge in Philadelphia, leaving it perched precariously over a river.
21 January 2014	Other	A drilling company in Pennsylvania is found guilty for having an improperly lined pit at a well site.
22 January 2014	Spill/Dump	A pipeline in Alberta ruptured and spilled 422,675 gallons (1.6 million litres) of waste fluids.
23 January 2014	Explosion	An Ohio gas well exploded in a fireball 100 feet (30.4m) high. Workers managed to evacuate the site immediately before the explosion after hearing a strange noise.
24 January 2014	Explosion	A tanker driver was killed when it exploded in Arkansas. A build up of methane from the waste fluids was suspected of causing the event. People more than a mile (1.6km) away reported feeling the explosion.
25 January 2014	Explosion	A pipeline near Winnipeg exploded, sending flames over 656 feet (200m) high.
27 January 2014	Explosion	A battery facility powering an oilfield exploded in Oklahoma.
27 January 2014	Spill/Dump	1,000 gallons (3,785 litres) of oil spilled into the Delaware River in Pennsylvania.

31 January 2014	Spill/Dump	An oil company was fined $1 million for allowing and independent contractor to illegally dump 4,700 gallons (17,791 litres) of oil into a river.
31 January 2014	Spill/Dump	A train carrying fuel and hazardous material for the oil and gas industry derailed and spilled its contents in Mississippi, forcing people to evacuate their homes.
11 February 2014	Explosion	A gas well in Pennsylvania exploded, injuring one worker and killed another. The fire burned for five days before it could be brought under control. Chevron, the site operator, was confirmed as giving out gift vouchers for a pizza and soft drink – valued at $12 – to around 100 homes near the site of the incident.
12 February 2014	Spill/Dump	Around 35 barrels (5,565 litres) of oil spilled from a site in North Dakota.
12 February 2014	Spill/Dump	Oil sprayed over agricultural land next to a well site in North Dakota.
13 February 2014	Blowout	Up to 3,000 gallons (11,356 litres) of fracking Fluid from an oil well in North Dakota leaked per day after the blowout preventer failed. Some of the fluid entered a nearby stream.
13 February 2014	Spill/Leak	More than 600 barrels (95,393 litres) of waste fluids were spilled at a well site in North Dakota.
13 February 2014	Spill/Dump	1,000 gallons (3,785 litres) of oil spilled when a train crashed in Pennsylvania.

13 February 2014	Explosion	A gas pipeline exploded in Kentucky, injuring two people and destroying two houses. A nearby barn and several cars are set alight from the explosion and a 60-foot (18.2m) wide crater was formed. 20 homes were forced to evacuate.
18 February 2014	Other	A drilling company was found guilty of having improper casing to protect fresh groundwater on a site in Pennsylvania.
26 February 2014	Blowout	A gas well blowout in Los Angeles forced 40 people to evacuate their homes within one and a half mile (2.4km) radius. It took several days to cap the well.
4 March 2014	Explosion	A massive explosion occurred at a well site in Colorado, causing injuries and affecting nearby residents.
10 March 2014	Spill/Dump	A faulty wellhead in Pennsylvania sprayed oil, fracking fluids and waste into the surroundings.
11 March 2014	Spill/Dump	Over 200 industrial garbage bags containing radioactive filters used in hydraulic fracturing were found dumped at a petrol station.
11 March 2014	Spill/Dump	A drilling company was found guilty of failing to properly dispose of industrial waste and prevent pollution of watercourses in Pennsylvania.
13 March 2014	Pipeline/Infrastructure	Around 30 gallons (114 litres) of toxic fluid spilled from a pipeline in Pennsylvania linking a compressor station to a well pad after a build up of ice.

17 March 2014	Pipeline/Infrastructure	A 5-inch (1.7cm) long crack is found in a pipeline passing through a nature reserve in Ohio. More than 20,000 gallons (75,708 litres) of oil was estimated to have leaked before the discovery.
20 March 2014	Spill/Dump	A drilling company in Pennsylvania is found to have illegally discharged industrial waste into nearby watercourses.
21 March 2014	Other	A drilling company was found guilty of failing to properly store, transport, process or dispose of waste in Pennsylvania.
22 March 2014	Spill/Dump	A 4-mile (6.4km) wide oil spill was discovered by hikers in Utah.
25 March 2014	Other	A drilling company was found guilty of failing to properly dispose of industrial waste and prevent pollution of nearby watercourses in Pennsylvania.
25 March 2014	Traffic	A tanker rolled over in Ohio, spilling its load of drilling mud.
25 March 2014	Spill/Dump	At least 1,638 gallons (6,201 litres) of tar sands oil leaked from a BP refinery into Lake Michigan.
31 March 2014	Explosion	A pipeline facility in Washington exploded and caught fire, forcing up to 1,000 people to evacuate their homes.
1 April 2014	Spill/Dump	An unknown quantity of fracking fluid spilled from a containment area due to heavy rain, killing vegetation in the affected area.
2 April 2014	Traffic	A lorry carrying hazardous material rolled over in West Virginia, spilling a toxic liquid onto the road, closing it for 12 hours/

4 April 2014	Explosion	A Compressed Natural Gas vehicle exploded, killing the driver and injuring the passenger. It is thought the load shifted during a turn, causing the fuel system to compromise and explode.
5 April 2014	Other	A worker is crushed to death by a piece of falling equipment at a well site in Ohio.
5 April 2014	Explosion	A pipeline exploded, causing six nearby families to evacuate their homes. A ground movement was suspected of causing the event.
7 April 2014	Spill/Dump	A fracking filter was found illegally dumped beside a road in North Dakota.
9 April 2014	Spill/Dump	Around 650 barrels (103,342 litres) of oil and 450 barrels (71,544 litres) of fracking fluid spilled from a storage tank due to pressure build up at a site in North Dakota.
9 April 2014	Other	A substance used to add odour to natural gas leaked from a dehydration station in Pennsylvania.
18 April 2014	Spill/Dump	500 tons of soil was removed from a fluid storage pond after a large leak was discovered.
22 April 2014	Other	A Jury in Dallas, Texas awards a family nearly $3 million in compensation from Aruba Petroleum Inc. after the company's fracking operation were found to have been responsible for causing years of illness, deaths of pets and livestock and making the property uninhabitable for months at a time.

Date	Type	Description
23 April 2014	Explosion	The town of Opal in Wyoming was evacuated following an explosion at a natural gas processing plant.
23 April 2014	Spill/Dump	Around 20 bags of illegally dumped fracking filters were found in North Dakota.
25 April 2014	Other	Four workers were injured after lightning struck a drilling site in Arkansas.
27 April 2014	Other	A 20-year-old worker died at a well site in North Dakota of suspected hydrogen sulphide exposure.
30 April 2014	Explosion	Two workers were killed and nine others injured in an explosion at a Texas exploration and production site. The workers were all operating a rig separating the produced fluids.
30 April 2014	Explosion	A train carrying oil from fracking sites derailed and exploded into flames in Virginia, spilling oil into the James River and forcing hundreds of people to evacuate the area.
30 April 2014	Pipeline/Infrastructure	A pipeline in Alberta ruptured and spilled 1,000 litres (264 gallons) of oil and 9,000 litres (2,378 gallons) of produced fluids during maintenance works. An unknown quantity entered a nearby creek.
1 May 2014	Pipeline/Infrastructure	Around 50,000 litres (13,209 gallons) of oil and produced fluids spilled from a pipeline in Alberta just 2km (1.2 miles) away from a similar incident on the 30th April 2014, during maintenance works. Some of the fluids entered a nearby creek.

June 2014	Explosion	One person suffered injuries from an explosion at a gas compressor station in Wharton County, Texas. The flames reached up to 46 metres (150 feet) high and destroyed a section of the nearby road. A lorry was also engulfed in the flames, which was left to extinguish itself.
28 June 2014	Explosion	A well pad in Monroe County, Ohio was engulfed in flames after a vehicle blaze spread. 25 homes were evacuated.
4 July 2014	Other	A lawyer representing three families living near a fracking wastewater pit and well site in Pennsylvania claimed that all had been without access to their own water supplies for over a year since an investigation had found that one of the homes had suffered contamination of their water well following leaks from the pit since 2012.
10 July 2014	Pipeline/Infrastructure	An oil pipeline leaked 3.78 million litres (1 million gallons) of drilling fluids since the 4th July onto the ground at the Fort Berthold Indian Reservation in North Dakota, which had also seeped into a nearby bay that supplied the Reservation's drinking water and killed grasses, trees and bushes in the vicinity. Environmental cleanup crews estimated that it would take several weeks to clear the spill.

26 July 2014	Explosion	A fracking well site in Clarington, Ohio, erupted into a fire that lasted for a week before it could be extinguished. During that time, around 30 separate explosions occurred which scattered material over the surrounding area. Twenty lorries were engulfed in the flames and a vast quantity of fluids and chemicals numbering in the tens of thousands of gallons spilled from the site and into a nearby creek after mixing with the fire fighting water runoff. Around 70,000 fish were killed as far as 8km (5 miles) downstream. At the time of writing, local residents still remain uninformed about what chemicals were released into the air and water, or contaminated the soot that fell onto residential areas of the town, due to the State's fracking disclosure law.
28 July 2014	Spill/Dump	The largest fracking-related spill in Oklahoma State saw 480 barrels (76,314 litres) of hydrochloric acid leak from a storage tank and across an alfalfa field.
7 August 2014	Other	The Pennsylvania Department of Environmental Protection fined a fracking company in Washington County, Pennsylvania, for a leaking wastewater pit.
15 August 2014	Explosion	Two condenser tanks at a gas well in Pennsylvania exploded into fire for unknown reasons.
19 August 2014	Other	The Pennsylvania Department of Environmental Protection fined a gas drilling company more than $76,500 for a 27-hour long gas leak in early January following an error resulting in a damaged valve.

21 August 2014	Pipeline/Infrastructure	A natural gas pipeline in Garvin County caught fire and injured four workers during maintenance works.
26 August 2014	Other	The Pennsylvania Department of Environmental Protection issued a fine of $250,000 to the President of three oil and gas companies for unauthorised discharge of fluids, failure to report the discharges and failure to plug abandoned wells between 2010 and June 2014.
29 August 2014	Pipeline/Infrastructure	More than 300,000 litres (79,252 gallons) of oil and produced fluids spilled from a pipeline in Alberta and into a nearby stream.
1 September 2014	Blowout	An oil well in Tyler County, Texas, suffered a blowout after drilling into a pocket of high-pressure natural gas that blew off the well head and spewed oil and gas into the air and a nearby creek for 12 hours. A flight restriction on flying below 914 metres (3,000 feet) within a 8km (5 mile) radius was enforced until the 8th September. The stream of oil and gas did not stop until the 4th September, once the pressure in the well had fallen.
9 September 2014	Spill/Dump	The Pennsylvania Department of Environmental Protection issued a Notice of Violation to a fracking company after confirming that liquid from a wastewater pit had been leaking following elevated chloride levels were first detected on 11th July.

13 September 2014	Other	A pipeline worker in Louisiana was killed in unknown circumstances during maintenance on an offshore pipeline and two others were injured.
25 September 2014	Pipeline/Infrastructure	An unknown amount of a refined oil product leaked near Marion, Texas and a sheen was observed on a nearby tributary.
7 October 2014	Spill/Dump	The Pennsylvania Department of Environmental Protection announced that it was pursuing a $4.5 million fine on a gas drilling company after claiming that more than 200 holes had been found in the lining of a fracking wastewater pit which had leaked contaminants into the environment.
10 October 2014	Spill/Dump	1,000 barrels (158,987 litres) of fracking brine spilled from a corroded pipe at a well site near Arnegard, North Dakota. An unknown quantity polluted surrounding land and a nearby stream.
15 October 2014	Pipeline/Infrastructure	More than 52,000 litres (13,737 gallons) of oil and fluids were spilled from a pipeline near Slave Lake in Alberta.
17 October 2014	Spill/Dump	More than 300 barrels (47,696 litres) of oil and contaminated water and an unknown quantity of mist leaked from a well near Watford City, North Dakota. It was still reported as leaking uncontrollably the following day.
19 October 2014	Pipeline/Infrastructure	Around 4,000 barrels (635,949 litres) of oil spilled from a pipeline in Louisiana and killed a number of fish and other small animals.

Date	Type	Description
21 October 2014	Other	An oil rig worker near Parachute, Colorado was killed by a high-pressure valve that was blown off a leaking standpipe and struck him on the back of the head. Once the drilling fluids had drained away, his body was recovered 1.8 metres (6 feet) away from where he had been standing at the time of the incident.
25 October 2014	Explosion	An explosion at an Ohio fracking well hospitalised a worker with serious burns.
28 October 2014	Explosion	An underground pipeline in Monroe County, Ohio exploded in a massive fireball in the early hours of the morning. It was not cleared until later that evening. Several acres of woodland were ignited in the incident.
29 October 2014	Other	A mandatory 3.2km (2 mile) radius evacuation zone was declared around a gas well in Jefferson County after the crew had accidentally sheered off the well head and released vast quantities of methane and other natural gasses.
30 October 2014	Blowout	400 families living near a fracking well that suffered a blowout on the 28th October were not permitted to return to their homes for two days.
12 November 2014	Pipeline/Infrastructure	More than 500,000 litres (132,086 gallons) of produced fluids spilled from a pipeline near Vauxhall, Alberta.
13 November 2014	Explosion	A worker was killed in an explosion and ensuing fire at an oil and gas well site in Ohio. The incident was thought to be the result of a condenser tank catching fire.

Date	Type	Description
13 November 2014	Blowout	A worker was killed and two others injured when a frozen high pressure water line ruptured as crews attempted to thaw it at a fracking well in Weld County, Colorado.
15 November 2014	Other	Four workers were killed and a fifth worker hospitalised following a leak of toxic methyl mercaptan gas at a chemical plant near Houston, Texas. The chemical is added to odourless natural gas.
17 November 2014	Explosion	An explosion at an oil field in New Mexico killed one worker and critically injured two others who were airlifted to hospital. A fourth worker was also injured and taken to another hospital.
21 November 2014	Explosion	A compressor station in Oklahoma exploded for unknown reasons. Two workers were set alight and suffered critical injuries from the explosion whilst the nearby school was evacuated.
22 November 2014	Spill/Dump	A report by the New York Times uncovered the fact that in October 2014, 14.38 million litres (3.8 million gallons) of fracking fluids had been spilled in North Dakota alone, and that since fracking commenced in that State in 2006, 69.65 million litres (18.4 million gallons) of toxic fluids and another 19.68 million litres (5.2 million gallons) of non-toxic fluids had been spilled or leaked into the environment.

28 November 2014	Pipeline/Infrastructure	The Ohio State Environmental Protection Agency fined a drilling company more than $300,000 for 19 counts of spillages in eastern Ohio between September 2012 and November 2013.
29 November 2014	Pipeline/Infrastructure	Up to 60,000 litres (15,850 gallons) of oil spilled from a pipeline in Alberta, Canada.
1 December 2014	Explosion	A Compressor Station in Susquehanna County, Pennsylvania exploded in the early morning.
3 December 2014	Other	A drilling company was charged with "Inadequate, insufficient, and/or improperly installed cement" in casing at a well in Venango County, Pennsylvania.
3 December 2014	Other	A drilling company in Susquehanna County, Pennsylvania was fined $120,000 (£78,290) for an explosion at a well site in January that injured a worker and spilled 10,732 litres (2,835 gallons) of produced fluids. The explosion was caused by a worker inspecting the inside of a produced fluids storage tank using a mobile phone for a light source.
4 December 2014	Other	A drilling company in Bradford County, Pennsylvania was charged with "Inadequate, insufficient, and/or improperly installed cement" and "Failure to prevent migration of gas or fluids into sources of fresh water causing pollution or diminution".
11 December 2014	Traffic	The driver of a water tanker serving oil and gas wells was killed after the vehicle slid on ice in Susquehanna County, Pennsylvania.

11 December 2014	Blowout	A large gas leak at a well in Burleson County, Texas forced the evacuation of around 30 homes within a 3.2km (2 mile) radius.
14 December 2014	Blowout	Thirty homes were evacuated with 2.4km (1.5 miles) of a well pad in Monroe County, Ohio. Specialist crews were unable to regain control of the leak until the 24th December, when the affected residents were finally allowed to return to their homes.
22 December 2014	Other	A drilling company was fined $2.3 million (£1.5 million) and told to spend an estimated $3 million (£1.95 million) to restore eight sites damaged by unauthorised discharge of hydraulic fracturing materials into streams and wetlands in West Virginia.
22 December 2014	Other	A pipeline company was fined $800,000 (£521,937) for violations in four different counties in Pennsylvania, including the discharge of sediment into protected waterways.
23 December 2014	Other	A Colorado-based drilling company was fines nearly $1 million (£652,422) for damage to streams as a result of a landslide and wastewater spill at a well pad in Greene County, Pennsylvania.
24 December 2014	Explosion	Around 12 homes were forced to evacuate on Christmas Eve after and explosion and fire at a monitoring station in Washington County, Pennsylvania.
25 December 2014	Other	Testing undertaken by Duke University revealed that contaminants associated with oil and gas wastewater had migrated into a creek upstream of a drinking water abstraction pipe.

30 December 2014	Other	Scientists from the University of Michigan and NASA verified the largest methane leak ever caused in the USA in the Four Corners region. The total cost of methane leaks from oil and gas drilling sites across the country was estimated at $2 billion (£1.3 billion).

GLOSSARY

Annulus — The space between the outside of the steel casing and the inside of the borehole. It is filled with cement during the casing process

Aquifer — Layers of porous rock close to the surface that contain fresh water and are a source of drinking water

Blowout Preventer — A safety device that connects the drilling rig and later equipment to the well bore

Borehole — A hole in the ground created by a drill

Casing — Layers of steel tubing and cement that are inserted into the borehole to prevent contamination of the surrounding geology by fluids inserted into and extracted from the well

Casing Shoe — An opening at the end of each section of casing that allows liquid cement to be pumped into the well and forced up through the annulus

Cellar — A large pit created in a well pad over which the drilling rig is located and through which the borehole is drilled. It is designed to trap any leaks originating from the wellhead and any spillages or leaks from the drilling equipment

Christmas Tree	A piece of equipment attached to the blowout preventer once the rig is removed through which fluids can be injected into and removed from the well
Conductor Pipe	The first piece of casing that is inserted into the ground through the base of the cellar. It provides a foundation to the surface equipment and acts as an earth to prevent a build up of static electricity
DECC	Department for Energy and Climate Change. The main Government department responsible for overseeing oil and gas drilling in the UK
Drill Bit	The tool used to cut through rock during drilling
Drill Mud	The fluid injected into the borehole during drilling to act as a coolant, sealant and a means of bringing the cuttings to the surface for collection and disposal
Drill Pipe	The piece of equipment on which the drill bit is attached and connected to the rig
Fracking Fluid	The fluid used in fracturing shale rock and other tight formations once drilling is complete. It typically consists of water, a propant and chemical additives
Fugitive Methane	Methane gas that unintentionally escapes into the atmosphere from drilling sites, for example through incomplete flaring

Heel	The point in a well bore just before where a section of perforation and fracturing has taken place
Hydrocarbons	The chemical name for oil- and gas-based fossil fuels. It references the two chemical elements that fossil fuels are made from that are a source of energy – hydrogen and carbon
Kick Off Point	The point in the drilling process where the curve leading to a horizontal leg begins to extend out from a vertical well
Lateral Well	The technical term for a horizontal well
Material Safety Data Sheet	A document produced for chemical substances that details the physical properties of a named substance, health and environmental hazards, emergency procedures and a range of other important data
Minerals Planning Authority	The local body responsible for determining applications for minerals exploration and extraction, such as a County Council
NORMs	Naturally Occurring Radioactive Materials. Radioactive materials that occur naturally deep underground such as Radon and Radium that are brought up in produced fluids or, in the case of gasses, flared off.

PEDL	Petroleum Exploration and Development Licence. An area of land that a particular company is permitted to explore for oil and gas
Perforating Gun	The equipment containing shaped explosive charges lowered into the well that is used to blow dozens of holes in casing to allow oil and gas to enter the well.
Produced Fluids	The liquid that is brought to the surface during and after drilling. It contains a mix of drilling fluids, fracking fluids, oil, gas and any other naturally occurring material released from the rocks
Production Casing	The technical name for the casing inserted through the Conductor Pipe and into the borehole
Propant	A solid ingredient used in fracking fluid to hold open the artificial fractures created in the rock. Silica sand is typically used
Protectors	The term used in the anti-fracking community to describe those taking part in community protection camps and blockades and is preferred to the term 'protestor'. It signifies a person committed to defending a community against fracking using non-violent and peaceful means

Scoping Opinion A process by which a developer can ask a planning authority for their opinion on what information should be supplied in an Environmental Statement before an application is made. It typically provides the developer with information on what the planning authority consider to be the main impacts of a proposed development that should be addressed by the applicant

Seismic Survey A method of using sound waves to create an image of the underground geology at a given point, much like an ultrasound is used during a pregnancy. The sounds can either be created via buried explosive charges or via special vehicles that can induce sound waves by vibrating a pad on the surface of the ground

Shale Rock The generic term for a type of oil or gas-bearing rock, usually buried deeply underground, that is not porous and cannot be exploited using conventional techniques

Surface Casing The first section of well casing inserted through the Conductor Pipe and through which all other production casing is inserted

Syngas Short for Synthetic Gas or Synthesis Gas. The name for the artificial gas produced during underground coal gasification

Tight Formation	The generic term for shale and non-shale rock that is not porous and requires hydraulic fracturing or other unconventional stimulation
Tripping Pipe	The technical term for the periodic removal of the drill bit and pipe from a borehole to allow for tool changes
Well Head	The term for the visible surface equipment connected to a well, such as the blowout preventer and Christmas tree
Well Pad	The term for a self-contained site prepared and used to commence a drilling operation.
Workover Rig	The second type of rig assembled over a borehole once the drill rig has completed its work. It is usually smaller than the drill rig and is used to insert and remove equipment such as the perforating gun into and out of a well

Printed in Great Britain
by Amazon.co.uk, Ltd.,
Marston Gate.